NEBOSH HEALTH AND SAFETY MANAGEMENT FOR CONSTRUCTION (UK)

Unit CN1 - Part 2

Element 8: Musculoskeletal Health and Load Handling

Element 9: Work Equipment

Element 10: Electricity

Element 11: Fire

Element 12: Chemical and Biological Agents

Element 13: Physical and Psychological Health

Contributors

Kevin Coley, MSc, BA, CMIOSH

With thanks to:

Mr Roger Passey, CMIOSH, MIIRSM

Dr T Robson, Bsc (Hons), PhD, CFIOSH, MRSC, CChem

Katharine Massey, GradIOSH, DipNEBOSH, MIIRSM

© RRC International

All rights reserved. RRC International is the trading name of The Rapid Results College Limited, Tuition House, 27-37 St George's Road, London, SW19 4DS, UK.

These materials are provided under licence from The Rapid Results College Limited. No part of this publication may be reproduced, stored in a retrieval system, or transmitted in any form, or by any means, electronic, electrostatic, mechanical, photocopied or otherwise, without the express permission in writing from RRC Publishing.

For information on all RRC publications and training courses, visit: www.rrc.co.uk

RRC: Unit CN1 Part 2

ISBN for this volume: 978-1-912652-63-1
Second edition February 2023

ACKNOWLEDGMENTS

RRC International would like to thank the National Examination Board in Occupational Safety and Health (NEBOSH) for their co-operation in allowing us to reproduce extracts from their syllabus guides.

This publication contains public sector information published by the Health and Safety Executive and licensed under the Open Government Licence v.3 (www.nationalarchives.gov.uk/doc/open-government-licence/version/3).

Every effort has been made to trace copyright material and obtain permission to reproduce it. If there are any errors or omissions, RRC would welcome notification so that corrections may be incorporated in future reprints or editions of this material.

Whilst the information in this book is believed to be true and accurate at the date of going to press, neither the author nor the publisher can accept any legal responsibility or liability for any errors or omissions that may be made.

Contents

Health and Safety Management for Construction (UK)

Element 8: Musculoskeletal Health and Load Handling

Musculoskeletal Disorders and Work-Related Upper Limb Disorders	8-3
Introduction to Musculoskeletal Disorders	8-3
Meaning of Terms	8-3
Examples of Repetitive Construction Activities that Can Cause MSDs and WRULDs	8-4
Possible Ill-Health Conditions from Poorly Designed Tasks and Workstations	8-4
Avoiding/Minimising Risks from Poorly Designed Tasks and Workstations	8-5

Manual Handling Hazards and Control Measures	8-7
Introduction to Manual Handling Hazards	8-7
Common Types of Manual Handling Injuries	8-7
Good Handling Technique For Manually Lifting Loads	8-9
Assessment of Manual Handling Risks	8-10

Load-Handling Equipment	8-17
Introduction to Lifting and Moving Equipment	8-17
Hazards and Controls For Common Types of Load-Handling Aids and Equipment	8-18
Requirements for Lifting Operations	8-27
Periodic Inspection and Examination/Testing of Lifting Equipment	8-27

Summary	8-29

Exam Skills	8-30

Contents

Element 9: Work Equipment

General Requirements for Work Equipment	**9-3**
Introduction to the General Requirements for Work Equipment	9-3
Scope of Work Equipment	9-3
Providing Suitable Work Equipment	9-4
Preventing Access to Dangerous Parts of Machinery	9-6
When the Use and Maintenance of Equipment with Specific Risk Needs to be Restricted	9-16
Providing Information, Instruction and Training about Specific Risks	9-16
Why Equipment Should be Maintained and Maintenance Conducted Safely	9-16
Emergency Operations Controls	9-18
Hand-Held Tools	**9-21**
Considerations for Selecting Hand-Held Tools	9-21
Hazards of a Range of Hand-Held Tools	9-22
Machinery Hazards and Control Measures	**9-26**
Consequences as a Result of Contact with Hazards Identified in ISO 12100:2010	9-26
Hazards and Controls of a Range of Site Equipment	9-30
Working Near Water	**9-36**
Additional Appropriate Control Measures	9-36
Summary	**9-40**
Exam Skills	**9-41**

Element 10: Electricity

Hazards and Risks — 10-3
Risks of Electricity — 10-3

Control Measures — 10-8
Protection of Conductors — 10-8
Strength and Capability of Equipment — 10-8
Protective Systems - Advantages and Limitations — 10-8
Use of Competent People — 10-10
Use of Safe Systems of Work — 10-11
Emergency Procedures — 10-13
Inspection and Maintenance Strategies — 10-13

Control Measures for Working Underneath or Near Overhead Power Lines — 10-18
Legal Requirements for Working Near Power Lines — 10-18
Preventing Line Contact Accidents through Management, Planning and Consultation — 10-18
Use of Barriers to Establish a Safety Zone When Working Near Overhead Lines — 10-19
Means of Safely Passing Underneath Overhead Lines — 10-19
Key Emergency Procedures for Contact with an Overhead Line — 10-20

Control Measures for Working Near Underground Power Cables — 10-22
Planning the Work — 10-22
Using Cable Plans — 10-22
Use of Service Locating Devices — 10-23
Safe Digging Practices — 10-23
Use of Appropriate Tools, Locating Devices and Route Planning When Undertaking Excavation Work — 10-24

Summary — 10-25

Exam Skills — 10-26

Contents

Element 11: Fire

Fire Principles	**11-3**
Basic Principles of Fire	11-3
Classification of Fires and Electrical Fires	11-4
Basic Principles of Heat Transmission and Fire Spread	11-4
Common Causes and Consequences of Fires within the Construction Industry	11-5
Preventing Fire and Spread	**11-8**
Control Measures to Minimise the Risk of Fire Starting in a Construction Workplace	11-8
Fire Alarms and Fire-Fighting	**11-14**
Common Fire Detection and Alarm Systems	11-14
Portable Fire-Fighting Equipment	11-15
Extinguishing Media	11-17
Access for Fire and Rescue Services and Vehicles	11-18
Summary	**11-20**
Exam Skills	**11-21**

Element 12: Chemical and Biological Agents

Hazardous Substances	12-3
Introduction to Forms, Classification and the Health Risks from Hazardous Substances	12-3
Forms of Chemical Agent	12-3
Forms of Biological Agents	12-4
Health Hazards Classifications	12-4

Assessment of Health Risks	12-7
Routes of Entry	12-7
What Needs to be Taken into Account When Assessing Health Risks	12-10
Sources of Information	12-11
Limitations of Information Used When Assessing Risks to Health	12-13
Role and Limitations of Hazardous Substance Monitoring	12-14
Purpose of Occupational Exposure Limits and How They Are Used	12-18

Control Measures	12-22
The Need to Prevent Exposure	12-22
Adequately Control Exposure	12-22
Principles of Good Practice	12-23
Common Measures Used to Implement the Principles of Good Practice	12-23
Additional Controls for Carcinogens, Asthmagens and Mutagens	12-33

Specific Agents	12-35
The Prevalence of Occupational Lung Disease Among Construction Workers	12-35
Health Risks, Controls and Likely Workplace Activities/Locations Where They Can be Found	12-36
Health Risks from and Controls for Working with Asbestos	12-40
Duty to Manage Asbestos	12-41

Summary	12-48

Exam Skills	12-49

Contents

Element 13: Physical and Psychological Health

Noise	**13-3**
Introduction to Noise	13-3
The Physical and Psychological Effects of Exposure to Noise	13-3
Commonly Used Terms in the Measurement of Sound	13-5
When Exposure Should be Assessed	13-5
Comparison of Measurements to Exposure Limits Established by Recognised Standards	13-6
Basic Noise Control Measures	13-7
Purpose, Application and Limitations of Personal Hearing Protection	13-8
Role of Health Surveillance	13-9
Vibration	**13-10**
The Effects on the Body of Exposure to Vibration	13-10
When Exposure Should be Assessed	13-11
Comparison of Measurements to Exposure Limits Established by Recognised Standards	13-13
Basic Vibration Control Measures	13-14
Role of Health Surveillance	13-15
Radiation	**13-16**
Differences Between Types of Radiation and their Health Effects	13-16
Typical Occupational Sources of Radiation	13-20
Basic Ways of Controlling Exposure to Radiation	13-21
Basic Radiation Protection Strategies	13-22
The Role of Monitoring and Health Surveillance	13-24
Mental Ill Health	**13-25**
The Frequency and Extent of Mental Ill Health in the Construction Industry	13-25
Recognising Common Symptoms	13-25
Causes of and Controls for Mental Ill Health	13-27
Recognition That Most People with Mental Ill Health Can Continue to Work Effectively	13-29
Organisations That Provide Support	13-29
Violence at Work	**13-31**
Introduction to Violence at Work	13-31
Types of Violence at Work	13-31
Effective Management of Violence at Work	13-33
Substance Abuse at Work	**13-35**
Risks to Health and Safety from Substance Abuse at Work	13-35
Managing Substance Abuse at Work	13-35
Summary	**13-38**
Exam Skills	**13-39**

Final Reminders

Suggested Answers to Study Questions

Element 8

Musculoskeletal Health and Load Handling

Learning Objectives

Once you've studied this element, you should be able to:

1. Explain work processes and practices that may contribute to musculoskeletal disorders, work-related upper limb disorders and the appropriate control measures.

2. Explain the hazards and control measures which should be considered when assessing risks from manual handling activities.

3. Explain the hazards and control measures to reduce the risk in the use of lifting and moving equipment with specific reference to manual and mechanically operated load moving equipment.

Contents

Musculoskeletal Disorders and Work-Related Upper Limb Disorders	**8-3**
Introduction to Musculoskeletal Disorders	8-3
Meaning of Terms	8-3
Examples of Repetitive Construction Activities that Can Cause MSDs and WRULDs	8-4
Possible Ill-Health Conditions from Poorly Designed Tasks and Workstations	8-4
Avoiding/Minimising Risks from Poorly Designed Tasks and Workstations	8-5
Manual Handling Hazards and Control Measures	**8-7**
Introduction to Manual Handling Hazards	8-7
Common Types of Manual Handling Injuries	8-7
Good Handling Technique For Manually Lifting Loads	8-9
Assessment of Manual Handling Risks	8-10
Load-Handling Equipment	**8-17**
Introduction to Lifting and Moving Equipment	8-17
Hazards and Controls For Common Types of Load-Handling Aids and Equipment	8-18
Requirements for Lifting Operations	8-27
Periodic Inspection and Examination/Testing of Lifting Equipment	8-27
Summary	**8-29**
Exam Skills	**8-30**

Musculoskeletal Disorders and Work-Related Upper Limb Disorders

IN THIS SECTION...

- Many musculoskeletal disorders and work-related upper limb disorders occur each year from manual handling and load moving activities in construction. There are common repetitive handling activities in construction work and common injuries that they can cause.
- Poorly designed tasks and workstations also contribute to ill-health effects.
- Factors contributing to ill-health conditions include the task, the working environment and the equipment used.
- Control measures include matching the workplace to individual needs of workers.

Introduction to Musculoskeletal Disorders

The following legislation may apply to avoiding musculoskeletal hazards from manual and mechanical handling operations in construction:

- The **Manual Handling Operations Regulations 1992 (as amended)**.
- The **Lifting Operations and Lifting Equipment Regulations 1998 (LOLER)**.
- The **Provision and Use of Work Equipment Regulations 1998 (PUWER)**.
- **Construction (Design and Management) Regulations 2015** (Regulation 9 - Duties of Designers).

TOPIC FOCUS

The HSE's guidance on the **Manual Handling Operations Regulations 1992** (L23, 4th edition, 2016) makes the 'relevant self-employed' person responsible for their own safety during manual handling. They should take the same steps to safeguard themselves as employers must to protect their employees, in similar circumstances. However, employers may be responsible for the health and safety of someone who is self-employed for tax and National Insurance purposes but who works under their control and direction.

Meaning of Terms

- **Musculoskeletal Disorders (MSDs)**

    A wide-ranging term that covers all disorders that affect the body's muscles, ligaments, tendons, joints, nerves and other soft tissues. Included are upper-limb disorders.

- **Work-Related Upper Limb Disorders (WRULDs)**

    Injuries occurring in the upper body (the hands, arms, wrists, fingers, neck and shoulders) usually affecting the soft tissue, and caused or contributed to by a worker's activities in the workplace.

- **Musculoskeletal Hazards**

    Often associated with lifting and moving heavy loads, but they are also associated with the way apparently easy and light objects are handled, and from poor posture whilst carrying out work activities.

8.1 Musculoskeletal Disorders and Work-Related Upper Limb Disorders

Examples of Repetitive Construction Activities that Can Cause MSDs and WRULDs

Musculoskeletal Disorders (MSDs) and Work-Related Upper Limb Disorders (WRULDs) are relatively common among construction workers. They generally occur as a result of repetitive handling activities - many construction activities fit this description, including:

- **Digging** - using a shovel by hand. This requires bending, twisting and lifting a load, often while stooping. If digging out a trench or other excavation, it may also involve turning and throwing the load long distances from the shovel up onto the surrounding ground of the trench, often using strenuous and jerky movements.
- **Kerb laying** - lifting heavy kerbstones, usually in very poor, stooped or even kneeling positions, and laying them accurately in-line and at the same level.
- **Movement and fixing of plasterboard** - requires the lifting (usually from a flat position to upright) and carrying of a very wide and tall plasterboard panel, then locating and fixing it in place. It often involves stretching.
- **Placement and finishing of concrete slabs** - usually lifting from a flat position, carrying and laying again in a flat position, before butting correctly to the other laid slabs and levelling. This involves a lot of work in poor posture, usually kneeling.
- **Bricklaying** - often involves a lot of repetition as the items are small, although not heavy. It requires a lot of twisting and turning movements to pick up and then lay the brick, while laying the cement requires similar repetition and movement. Laying can start at ground level and progress to reaching positions.
- **Erecting and dismantling scaffolds** - requires reaching to remove poles and components from a vehicle, turning and carrying to a location (often some distance), then handling, turning and twisting to locate and fix the pole. Dismantling and erecting can both involve work at height.
- **Use of display screen equipment** - architects and designers, as well as office staff and site management, may require the use of computers, and may be in temporary accommodation on site. This may not have the best heating, lighting, etc., and equipment may not always be suitable (chairs, tables, etc.). Poor posture is an important issue.

Using display screen equipment on site

Possible Ill-Health Conditions from Poorly Designed Tasks and Workstations

The ill-health effects due to poor design of tasks and workstations include:

- **Physical stress** - resulting in injury or general fatigue (aches, pains, etc.), usually from poor posture and excessive demands on manual dexterity, but also in respect of exposure to excessive noise and vibration.
- **Visual problems** - often through excessive brightness or prolonged, concentrated work on small objects, either on a computer display screen, or in respect of components used in a work process, e.g. in a construction site drawing office.
- **Mental stress** - mainly through excessive demands of task performance, lack of control over working processes and poor organisational and physical environmental conditions.

These effects are generally chronic ones, brought about by prolonged exposure to the activity or conditions.

Musculoskeletal Disorders

Typical problems from poorly designed tasks and workstations include:

- **Work-Related Upper Limb Disorders (WRULDs)**, such as:
  - Tendonitis - a swelling or irritation of a tendon which induces pain, tenderness or restricted movement of the muscle attached to the affected tendon.
  - Tenosynovitis - an inflammation of the sheath surrounding the tendon causing pain, tenderness and swelling over the tendon.
  - Ulnar neuritis - an inflammation of the ulnar nerve which runs down the full length of the arm to the hand and controls muscles that move the thumb and fingers. This causes a 'pins and needles' feeling and pain in the forearm and the fourth and fifth fingers.
  - Carpal tunnel syndrome - numbness, tingling and pain in the thumb, index and middle fingers, resulting from the compression of the median nerve which sends signals to and from the thumb and first three fingers.
  - Thoracic outlet syndrome - a compressive disorder of the nerve and blood vessel between the neck and the shoulder.
- **Back disorders**, including disc injuries, aches and pains.

Avoiding/Minimising Risks from Poorly Designed Tasks and Workstations

Various factors influence the risk of ill-health conditions from work activities. These relate to the **task**, the **environment** and the **work equipment** used.

Task

The key risk factors associated with ill-health conditions are:

- **Posture and physical action** - the position of the body and the way it has to move to carry out the task. Problems arise from stress on particular parts of the body caused by awkward posture and movements in the task. These are affected by the work methods (bending, twisting) and layout (space, high and low storage locations).
- **Forces involved** - how strenuous a particular activity is. The greater the forces involved, the greater the stress placed on the body and the greater the risk of injury. Forces are determined by the speed, degree of movement and the size of the object being used.
- **Repetition** - how often the physical action (lifting, twisting, bending, squeezing, etc.) has to be performed during a task. In general, the more repetitions there are over a short period of time, the greater the risk of injury.
- **Duration and recovery time** - the amount of time a worker takes to carry out a task (in relation to physical stress of position and movement) and the amount of time the body has to recover from any such stress. Recovery time may be:
  - Time spent undertaking other activities which do not involve similar stresses.
  - Time specifically set aside to allow the body to recover from being under physical stress.

8.1 Musculoskeletal Disorders and Work-Related Upper Limb Disorders

Environment

The key risk factors here are lighting and glare. Problems may arise in respect of:

- **Illumination** - the lighting level, which must be appropriate to the type of work being carried out, e.g. the finer the detail of the work, the higher the illumination required.
- **Contrast** - significant variations in lighting levels between different parts of the field of vision or different areas of the workplace. These can temporarily blind the worker or interfere with vision as the eye adjusts to the different levels.
- **Glare** - occurs when one part of the visual field is much brighter than the average brightness to which the visual system is adapted. Direct interference with vision is known as disability glare. Where glare causes discomfort, annoyance, irritability or distraction, this is known as discomfort glare and is often related to symptoms of visual fatigue such as sore eyes and headaches.

Moving between dark and bright areas can interfere with vision

  Glare may occur:

  - As a result of a light source being directly in the line of vision (e.g. the setting sun shining in through a vehicle windscreen, or a badly positioned floodlight on site).
  - From the reflection of light off a polished surface, e.g. a mirror or computer's screen.

In addition, much of the construction environment is outdoors, and cold or damp conditions can make it unsatisfactory for workers. This can lead to a lack of motivation among employees and possible increased lost time due to ill health, allied with a possible effect on the equipment used.

Equipment

The characteristics of the equipment used to carry out work activities can themselves increase the risk of harm by putting extra strain on the body in two main ways:

- The physical characteristics of the equipment itself, for example:
    - Difficult to manipulate (handles too small or too large to grip easily).
    - Encourages poor posture (non-adjustable seats).
- The position of the equipment in relation to the worker in the position they normally occupy, for example:
    - Bending over a conveyor belt used to put materials onto a roof, such as tiles, cement.
    - Having to reach down to pick up items from a low position, such as bricks off a pallet.

It is important, therefore, to consider the requirements of the equipment user, and the ways in which the positioning, layout, size, etc. of the equipment can be **adjusted** to suit the user. Some improvements could include:

- Easily gripped handles - good shape and size.
- Seats that adjust to suit the user.
- Conveyors, pallets, etc. put at suitable height to place or retrieve items.

STUDY QUESTIONS

1. What is a WRULD and how might it be brought about?
2. Outline three construction activities that can cause MSDs or WRULDs.
3. What are the three main ill-health effects of poorly designed tasks and workstations?
4. Environmental factors can contribute to ill-health conditions. Describe how glare can contribute, and identify common sources.

(Suggested Answers are at the end.)

Manual Handling Hazards and Control Measures

IN THIS SECTION...

- There are several types of manual handling injuries that can occur from construction activities, including: back injuries; musculoskeletal problems; hernias; cuts, abrasions and bruising; bone injuries; and WRULDs.
- Assessment of manual handling risks should consider the Task, Individual, Load and Environment ('TILE').
- Means of avoiding and minimising the risks from manual handling in construction activities can be considered by looking at the elements of 'TILE' and addressing each of these.
- There are recognised efficient movement principles for manually lifting loads to avoid MSDs due to lifting, poor posture and repetitive or awkward movements.

Introduction to Manual Handling Hazards

DEFINITIONS

MANUAL HANDLING

"...any transporting or supporting of a load (including the lifting, putting down, pushing, pulling, carrying or moving thereof) by hand or by bodily force."

Manual Handling Operations Regulations 1992 (as amended)

There are four main causes of harm from manual handling operations:

- Failing to use a proper technique for lifting and/or moving an object or load.
- Moving loads which are too heavy.
- Failing to grip an object or load in a safe manner.
- Not wearing appropriate PPE.

Harm from any of these actions may be an immediate injury (acute) or longer-term chronic injury and mobility problems.

Common Types of Manual Handling Injuries

The types of injury caused by handling objects and loads will depend on the part of the body which is put under stress by the action.

Moving heavy loads and failing to use a proper technique can result in injury

Back Injuries

Back injuries are very common in construction activities, they can be caused by twisting, lifting or pushing loads where the stress is borne on the spine, usually towards the base.

Excessive torsional or crushing movement on the spine leads to displacement of the discs (fluid-filled cushions between the vertebrae).

The most serious injuries are prolapsed or crushed discs, herniated discs and sciatica (a neuralgia of the hip and thigh from a trapped sciatic nerve).

More generalised debilitating back problems such as lumbago may also occur.

8.2 Manual Handling Hazards and Control Measures

Muscular Problems

These usually take the form of a strain or sprain:

- A strain is when a **muscle** is stretched beyond its normal limit **to the point where it may tear.**
- A sprain is where a **tendon** is subjected to sudden or excessive force **to the point where it may be twisted or torn**.

Either may tear or rupture the casing of the muscle, which is a serious injury. This causes weakened joints and restricted movement, making it painful.

They may be caused by stretching; lifting heavy loads; or slips, trips and falls. In most cases, these are acute injuries, but strains can build up over time.

Hernias

A hernia is a rupture in the musculature of the body cavity wall, usually in the lower abdomen, which allows a protrusion of part of the intestine. It is caused by excessive strain on abdominal muscles during lifting.

Cuts, Abrasions and Bruising

These will be caused by contact with:

- The surfaces of the objects being handled.
- Stationary objects while moving or placing a load.

Bone Injuries

Fractures and cracks are usually impact injuries caused by:

- Crushing part of the body - usually fingers - under a load.
- Dropping objects onto the feet.
- Slips, trips and falls.

Work-Related Upper Limb Disorders (WRULDs)

As was mentioned earlier in the element, upper-limb disorders affect the soft connecting tissues, muscles and nerves of the hand, wrist, arm and shoulder.

Severity may vary from occasional aches, pains and discomfort of the affected parts, through to well-defined and specific disease or injury. Loss of function may result in reduced work capacity.

WRULDs arise from ordinary movements, such as repetitive gripping, twisting, reaching or moving. Stress of movement or weight of load are not critical factors.

The effects are:

- General fatigue and loss of concentration and co-ordination.
- Inflammation of the tissue of the hand (elbow or knee) caused by constant bruising or friction.
- Temporary fatigue, stiffness or soreness of the muscles.
- Musculoskeletal disorders (outlined earlier in the element) including tendonitis, tenosynovitis, ulnar neuritis, carpal tunnel syndrome and thoracic outlet syndrome.

The key factors associated with the increased risk of WRULDs include:

- Excess force exerted to overcome resistance in a work operation due to poor design.
- Highly repetitive motions with short cycle times giving little time for recovery.
- Awkward postures causing significant stress to joints of the upper limbs and surrounding soft tissues.

Manual Handling Hazards and Control Measures | 8.2

Ill Health Due to Exposure to Hazardous Substances

Handlers may be exposed to hazardous substances such as cement, if bags break open while being handled.

Good Handling Technique For Manually Lifting Loads

There are recognised efficient movement principles (kinetic lifting techniques) for lifting loads - these should be used by all workers at all times to avoid WRULDs due to lifting and putting down loads, poor posture and repetitive or awkward movements.

Before Lifting

Before picking up a load, there are a number of checks to be made:

- Confirm the approximate or actual weight of the load.
- Check for awkward shapes, moving parts, etc.
- Plan the route and examine it for tripping and other hazards.
- Remove obstructions and clear work surfaces.
- Make sure you have somewhere to put the load down when you get there.
- Wear suitable clothing - PPE if necessary.
- Establish a firm grip when picking it up.

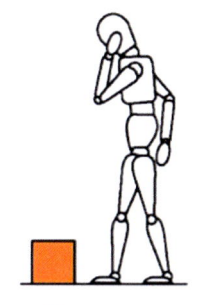

Preparation

The Lift or Movement of the Load

Good technique is essential here and includes the following:

- Adopt a position with one foot slightly in front of the other.
- Bend the knees.
- Keep the spine in an upright position.
- Avoid twisting, over-reaching and jerking.
- Establish a good balance.
- Keep the load close to the body at all times and maintain a firm grip.
- Use your body weight to lift the load or carry out a movement.
- Inhale before you pick it up.
- Exhale while you are picking it up.

Lifting

If a load is too heavy for one person to handle, seek assistance or use mechanical aids such as trolleys, pallet trucks, etc.

Putting Down or Completing the Task

The same good technique should be followed here as for picking up the load:

- Adopt a position with one foot slightly in front of the other.
- Keep the spine in an upright position.
- Bend the knees.
- Avoid twisting, over-reaching and jerking.
- Maintain a good balance.
- Keep the load close to the body at all times and maintain a firm grip.
- Use your body weight to manoeuvre and place the load.
- On completion, ensure the load is placed securely and does not create an obstruction.

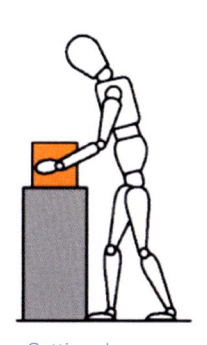

Setting down

8.2 Manual Handling Hazards and Control Measures

> **TOPIC FOCUS**
>
> **Manual Handling in Practice - Kerb Laying**
>
> Such work may involve a serious risk of injury. Kerbstones weigh around 67kg, the work is repetitive and encourages poor posture. Crushing fingers is common. The manual handling hierarchy should be applied:
>
> - **Eliminate** - by designing out the need to manually handle kerbstones.
> - **Substitute** - by using alternative (lighter) materials to make the kerbstones.
> - **Totally mechanise** - using machines to handle and lay the kerb.
> - **Partially mechanise** - do as much as possible using handling and laying equipment.
> - **Manually handle** - as a last resort, lay the kerb in short stretches, ensuring handlers are well trained and have all appropriate PPE. Use team lifts with the kerbstones.
>
> These elements of manual handling should be considered by all - the client, designers, principal contractor and competent person together with the manufacturers, to reduce the risks.

Assessment of Manual Handling Risks

The procedure for carrying out a manual handling risk assessment is illustrated in the following flowchart:

Manual Handling Hazards and Control Measures | 8.2

```
Does the work involve manual handling operations?
   │ YES                               └── NO ──→
   ▼
Is there a risk of Injury?
   │ YES/POSSIBLY                      └── NO ──→
   ▼
Is it reasonably practicable to avoid moving the loads?
   │ NO                                └── YES ──→
   ▼
Is it reasonably practicable to automate or mechanise the operations?  ── YES ──→  Does some risk of manual handling injury remain?  ── NO ──→
   │ NO                                                                              │ YES/POSSIBLY
   ▼                                                                                 │
Carry out manual handling risk assessment  ←─────────────────────────────────────────┘
   ▼
Determine measures to reduce risk of injury to the lowest level reasonably practicable
   ▼
Implement the measures
   ▼
Is the risk of injury sufficiently reduced?
   │ NO → (loop back to Determine measures)
   │ YES
   ▼
End of intital assessment – review if conditions change significantly
```

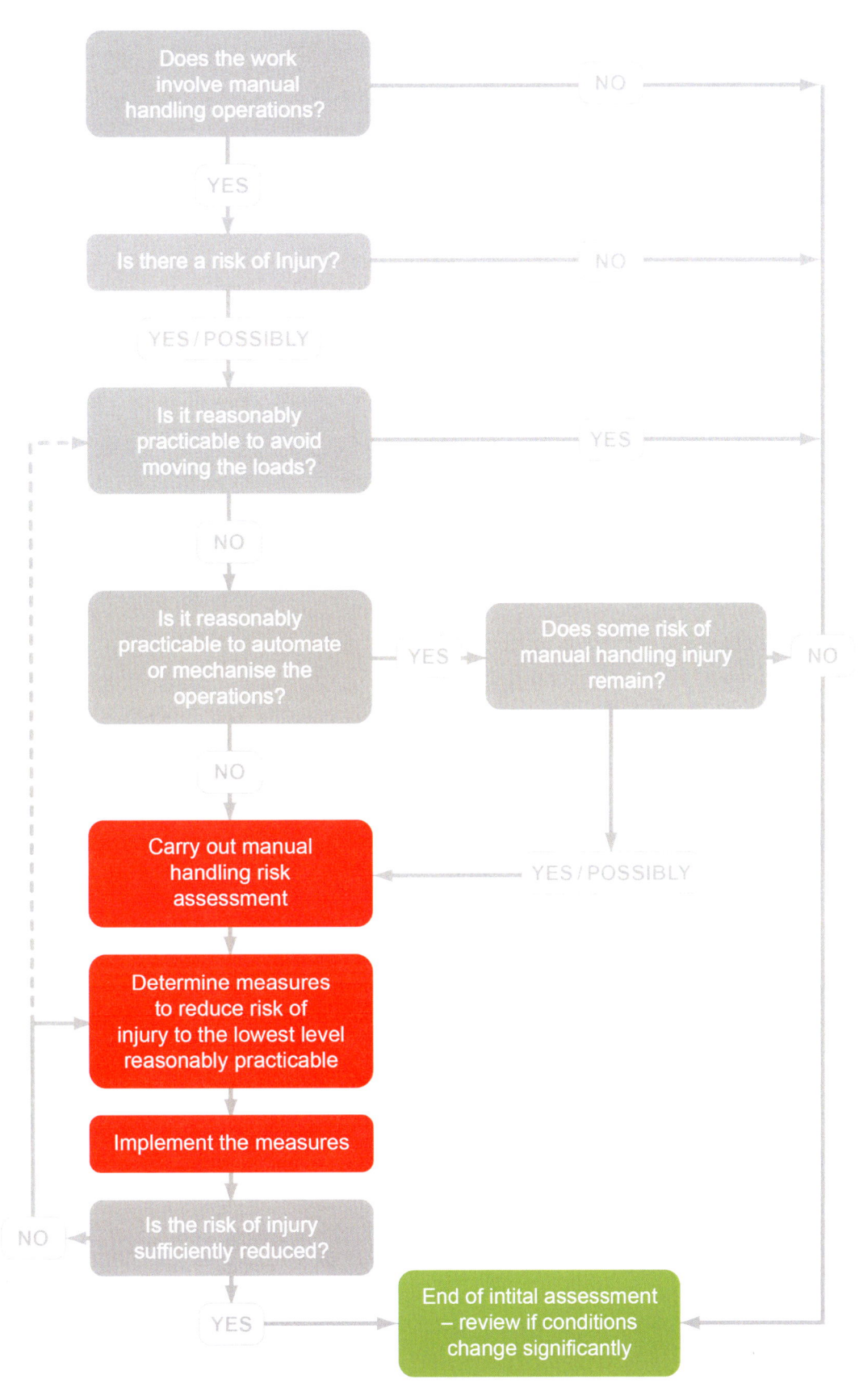

Manual handling risk assessment

8.2 Manual Handling Hazards and Control Measures

The identification of hazards in manual handling operations involves four key factors, as specified in Schedule 1 of the **Manual Handling Operations Regulations 1992 (as amended)**:

- The **task** - analysis of the nature of the handling operation and the identification of high-risk activities.
- The **load** - analysis, including measurements, of the object(s) being handled.
- The working **environment** - analysis of the immediate physical surroundings within which the handling operation takes place.
- **Individual capability** - consideration of the (mainly) physical characteristics of the person doing the handling operation and their ability in terms of knowledge and skills.

HINTS AND TIPS

A simple mnemonic to help you remember the four risk factors:

T	Task
I	Individual
L	Load
E	Environment

All tasks involving manual handling, however seemingly trivial, should be subject to a risk assessment. The findings should be recorded unless:

- it is simple and obvious; and
- the operation itself is low-risk and very brief.

Task

The types, frequencies and duration of movements should be analysed to identify those movements most likely to cause injury. Where there is a clear risk, or where the risk is uncertain, the task may be broken down into more detail, considering whether it involves aspects such as:

- Holding loads away from the trunk.
- Twisting.
- Stooping.
- Reaching upwards.
- Large vertical movement.
- Long carrying distances.
- Strenuous pushing or pulling.
- Unpredictable movement of loads.
- Repetitive handling.
- Insufficient rest or recovery.
- A work rate imposed by a process.

Movements, such as reaching upwards, should be analysed

Manual Handling Hazards and Control Measures | 8.2

Consideration should be given to the redesign of the task itself. The following aspects may be looked at:

- **Sequencing** - adjusting tasks to minimise the number of operations involving lifting or carrying loads.
- **Work routine** - reducing repetitive operations to allow variation in movement and posture by:
    - Introducing breaks.
    - Using job rotation.
    - Finding ways in which workers can operate at their own pace, rather than keeping up with a machine or process.
- **Using teams** - by specifying that two (or more) persons are required to lift certain types of load, or passing the load on (a 'chain') rather than carrying it all the way, particularly at changes in level.
- By **mechanising or automating the task** using:
    - Robots to carry out the work.
    - Automated assembly machines.
    - Mechanical handling equipment such as pallet trucks, powered trolleys, forklift trucks, etc.

Individual

Various characteristics of the individual can affect handling safety, such as attitude, lack of attention and whether appropriate PPE (e.g. gloves and safety footwear) are worn.

There are three main aspects to considering an individual's ability to carry out manual handling tasks safely:

- Does the task require unusual capabilities (e.g. strength or people of a particular height)?
    - Specific requirements should be made clear and only persons meeting them should carry out the task.
    - The level of risk in a task should be considered high if it could not be carried out by most reasonably fit employees.
- Does it present a risk to those with a health problem or to new or expectant mothers?
    - An individual's general health and fitness is a significant factor in their ability to undertake manual handling tasks.
    - It may be appropriate to carry out medical checks to establish a person's general fitness, or the implications of any relevant previous injury or complaint.
    - Specific assessment should address certain tasks. Lifting moderately heavy weights, bending and twisting, or standing for long periods may present particular risks during pregnancy and during the 12 weeks following a normal confinement.
- Does it require special information and/or training?
    - The assessment should consider the characteristics of the load or appropriate methods of carrying out the task. Specific training in working practices should be given where necessary.

All employees expected to carry out manual handling tasks must have adequate training, instruction, information and supervision to ensure they are capable of fulfilling these tasks at all times:

- Allowances should be made for those with particular health problems and new or expectant mothers, who may be more susceptible to injury.
- Adequate health monitoring and reporting systems must be in place.
- Training should be job-related and carried out in work conditions.

Employees should understand the importance of:

- The design of the tasks involved including the layout of the workplace.
- Recognising different types of loads.
- Assessing the weight and balance of loads.

8.2 Manual Handling Hazards and Control Measures

- Deciding which loads can be handled alone and which should be team-handled.
- Safe lifting and handling techniques - including the risks from careless and unskilled handling, e.g. holding loads away from the body will increase the risk of injury.
- The correct use of PPE.
- The correct use of mechanical aids - including power- and hand-operated devices.

Design characteristics of handling tasks and use of mechanical equipment can greatly improve manual handling on construction sites. Some manual handling solutions include:

- Using:
    - Single reusable adaptors to place road signs.
    - An inclined hoist and panel trolley to move and place sheets of plasterboard.
    - A sack truck to move and place kerbstones.
    - A valley gutter trolley to handle roofing materials.
    - An inclined hoist to load roof tiles.
    - A wheeled pallet truck to deliver bags of plaster, cement, etc.
- Placing:
    - Coping stones using a vacuum-powered handling device.
    - Concrete beams or lintels by using a hand-operated lift truck.
- Positioning roof trusses and bundles of scaffold tubes using a telehandler.

Load

> **DEFINITION**
>
> **LOAD**
>
> Defined as any discrete moveable object, including a person. The weight of the load is not the most critical factor.
>
> The term does **not** include a control lever attached to a machine.

- **Hazards**
    - Weight.
    - Size.
    - Shape.
    - Resistance to movement.
    - Rigidity or lack of it.
    - Position of centre of gravity.
    - Presence or absence of handles.
    - Surface texture.
- **Other Factors**
    - Any contents, e.g. cement (if container could break or come open during handling).
    - Stability of the contents.

Manual Handling Hazards and Control Measures | 8.2

Where there is a clear risk of injury, or where the risk is uncertain, the characteristics of the load may be broken down into more detail, considering aspects such as whether it is:

- Heavy.
- Bulky/unwieldy.
- Difficult to grasp.
- Unstable/unpredictable.
- Intrinsically harmful (e.g. sharp or hot).

The following aspects should be considered:

- Weight and size
    - Break up loads into lighter and/or smaller units. However, be aware that this may mean there are more elements to the task leading to repetition.
- Making the load easier to grasp
    - Where it is not possible to make the load smaller, hand grips or handles may be provided.
- Making the load more stable and rigid
    - Pack items within containers to ensure they do not move around inside, and that the weight is well distributed.
- Making the load less damaging to hold
    - Ensure that the surfaces of the load are clean, smooth, not slippery (from being wet or greasy, or even dusty) and, in the case of hot or cold items, that they are in insulated containers.
- Markings
    - Indicate approximate weight and centre of gravity, which way up to hold and stack the load, and warnings about instability.

Make loads more manageable to carry

Working Environment

The environment refers to the general and specific conditions in the immediate surroundings where the manual handling operations occur. This should include any routes taken by loads.

The key considerations are:

- Constraints on movement and posture, e.g. confined spaces, fixed chairs, or hindrance caused by certain types of clothing or PPE.
- Condition of floors (e.g. slippery, broken or uneven) and other surfaces (e.g. unstable shelving).
- Variations in levels - the presence of ramps, steps or ladders, shelving heights.
- Temperature and humidity:
    - High heat and humidity can cause dehydration and significantly increase the risks.
    - Extreme cold can make objects hazardous to touch and affect dexterity.
- Strong air movements (e.g. gusts of wind) may make loads unstable.
- Lighting conditions - poor general lighting and strong variations between light and shade can increase risks.

Ways to improve the conditions in which the manual handling is carried out:

- Workplace design:
    - Make access to the load being handled and any equipment being used as comfortable as possible.
    - Ensure adequate space for all movements.
    - Reduce the height over which a load is to be lifted and/or carried.

8.2 Manual Handling Hazards and Control Measures

- Floor conditions:
    - All floors should be free of obstructions, bumps, holes and any materials which may cause a worker to slip, trip, fall or otherwise lose their footing while undertaking the handling operation.
- Changes of level:
    - Avoid using steps and ladders when carrying loads. Ramps may need to be provided.
- Atmospheric conditions:
    - Heating and ventilation should ensure conditions are comfortable.
    - Ensure good lighting levels at all parts of the workplace
- Personal Protective Equipment (PPE):
    - Operators should have such PPE as is necessary to protect themselves from harm or damage as a consequence of the types of load handled.

STUDY QUESTIONS

5. What are the main causes of injury to persons as a result of manual handling operations?
6. What are the characteristics of a load which present a hazard?
7. Identify the main hazards presented by the working environment in respect of manual handling operations.
8. How can manual handling tasks be redesigned to make them less hazardous?

(Suggested Answers are at the end.)

Load-Handling Equipment

IN THIS SECTION...

- There are many hazards and control measures associated with the use of lifting and moving equipment. Such equipment includes:
  - Mechanically operated load-moving equipment, such as forklift trucks, telescopic handlers, hoists and cranes.
  - Manually operated load-moving equipment, such as sack trucks and pallet trucks.
- There are legal requirements for lifting operations, including:
  - Strong, stable and suitable equipment.
  - Correct positioning and installation.
  - Appropriate visible markings (e.g. safe working load).
  - Proper planning and supervision of lifting operations, including being carried out in a safe manner by competent people.
  - Special requirements for lifting equipment used for lifting people.
- There are also requirements for regular visual inspection and statutory requirements for the thorough examination and inspection of lifting equipment.

Introduction to Lifting and Moving Equipment

The **Lifting Operations and Lifting Equipment Regulations 1998 (LOLER)** applies to **all** lifting equipment used for work purposes, even where it was manufactured and put into use before **LOLER** came into force in 1998.

The following examples illustrate the type of equipment which can be used to aid lifting operations, and which should be assessed for the application of **LOLER** on construction sites:

- Cranes.
- Lift trucks and telescopic handlers.
- Hand pallet trucks, specifically those that have the ability to raise the forks.
- Goods lifts or passenger lifts.
- Simple systems, such as a rope and pulley used to raise a bucket of cement on a building site, a construction site hoist or a gin wheel.
- Pull-lifts.
- A scissor lift or articulated arm Mobile Elevating Work Platform (MEWP).
- Ropes used for climbing, work positioning or during structural examination of the external structure of a building.

8.3 Load-Handling Equipment

Hazards and Controls For Common Types of Load-Handling Aids and Equipment

Cranes

Tower cranes, gantry, overhead cranes on tracks and a variety of mobile cranes are used on construction sites.

The selection of the appropriate type of crane will depend on the:

- Weight and dimensions of the load.
- Heights of lifts and distances/areas of movement of loads.
- Number and frequency of lifts.
- Length of time the crane will be used.
- Site ground conditions.
- Space available for crane access, erection, operation and dismantling.
- Other special operational requirements (e.g. adjacent operations).

Mobile Cranes

These are the most common type of crane. Small units with rough terrain wheels and telescopic jib are useful on all construction sites. Others have rubber wheels or tracks.

Hazards include:

- Imbalance of the crane causing it to tip over.
- Overhead obstructions and power lines.
- Lack of stabilisers or outriggers during use.
- Overloading.
- High wind conditions.

Operational problems can arise for a range of reasons, such as:

- Jib swings out of control, striking a structure.
- Operator cannot fully see the load being handled.
- Crane is moved on soft or uneven ground.
- Load strikes something during horizontal movement.
- Crane is driven with a suspended load.
- Load, or part of it, falls.
- Operator error, or lack of competence, e.g. moving the load too quickly, or not having full visibility.
- Slinger/banksman error, e.g. not attaching the load correctly, or misguiding the operator.

Tower Cranes

These consist of a tall, slender mast with a jib at the top and are used in long-term construction activities with wide areas of operation.

Hazards include:

- Collapse of the tower due to incorrect construction.
- High wind conditions.
- Collapse or bending of the jib due to overloading or fatigue.
- Impaired view of the load for the operator.
- Swinging or unstable loads.
- Operator, or slinger error, or lack of competence.
- Operating outside the safe working radius.

Tower cranes are usually used in long-term construction activities

Because of the risks associated with the use of tower cranes and the potential for incorrect construction or failure to maintain the structures, regular inspection and maintenance is essential, as for all lifting equipment.

Accessories for Lifting Operations

A range of different accessories may be used in lifting operations, including:

- **Chains** - together with slings and strops - give a strong, flexible link between load and load hook, connected by shackles or D-links.
- **Two-, three- or four-legged slings** - steel wire ropes with eyes and hooks to connect to loads and the load hook.
- **Endless slings** - usually nylon or canvas webbing in a continuous loop. The sling is passed around a load, and each side of the loop is connected to the load hook.
- **Webbing straps (or strops)** - have a loop at each end to be connected to a load by a shackle at one end, and the load hook at the other.
- **Fibre ropes** - have many uses, including securing, slinging and guiding loads.
- **Shackles (or D-links)** - are passed through the eye of a sling or chain and a screw pin inserted to secure the connection.
- **Eyebolts** - threaded bolts with an eye formed at one end - used on machinery and items with a dedicated lifting point, and connected with a shackle or D-link.
- **Lifting beams (spreader beams)** - usually designed and made for specific lifting operations; made from steel with attachment points to connect to a load.
- **Hooks** - have a safety catch (or 'moused' with a knot) to which a sling, chain, strop, etc. is attached (the load hook) to be lifted.

Hazards arise if the lifting accessories are damaged (corroded, frayed, knotted, distorted, kinked, cracked, etc.) or used incorrectly (dragging wires and ropes across the ground, allowing ropes to slip when slack is taken up, stretching ropes or shackles, etc.) as they may fail, causing loads (materials or people) to fall from height, for example.

8.3 Load-Handling Equipment

Controls

> **TOPIC FOCUS**
>
> An appointed person will:
>
> - Control a lifting operation.
> - Be required to carry out an assessment of the operation, to check that:
>   - Ground conditions are suitable.
>   - There are no overhead obstructions.
>   - The lifting appliance is suitable with SWL clearly marked.
>   - Examination and test certificates are in place and up to date.
>   - Incident and defect reporting procedure is in place.
>   - Operators, slingers and banksmen are suitable and competent.
>   - A good method of communication (i.e. hand signals) is in place.
>   - The prevailing and expected weather conditions are suitable.

Capacity

The further a load is from the crane, the lower the maximum load it can lift, so all cranes have a load-radius chart to indicate these safe levels. Some safety devices will also help:

- Safe working load/radius indicators - show the SWL when the angle of the jib is varied and may operate automatically or provide read-outs for the crane operator.
- Automatic safe load indicators - give a visual warning to the crane operator when SWL is approached; and an audible warning if exceeded.

These must be tested before each operation.

An appointed person controls all lifting operations

Site Conditions

> **DEFINITION**
>
> **OVERSAILING**
>
> The action of carrying a suspended load over something, e.g. over other construction activities, buildings or roads/railways/waterways, etc.

All crane operations should be carried out on stable ground. Other key factors include:

- Access and egress routes must be of sound construction.
- Inclines should be eliminated or reduced to prevent overturning.
- Site on which the crane is located should be level and firm, with no holes or excavations close by to cause collapse.
- Site and lifting radius should be clear of overhead obstructions and nearby buildings/structures.
- No voids beneath the crane site, e.g. manholes, drains, culverts, or gas/water supply routes.
- Oversailing nearby properties or facilities such as railways should be prevented (or permission sought if required).

Load-Handling Equipment | 8.3

Competence

The lift should be planned and supervised by a competent person. Operators should be competent and experienced in the type of lifting equipment, and should have a clear view, or means of communication with, the slinger/banksman; they should inspect and take into account the ground conditions, any obstructions and weather factors. Slingers/banksmen should be competent in attaching the load, guiding the operator and using the recognised hand signals, or other means of communication.

Carrying Out the Lift

Key factors for preparation and lift include:

- Preventing unauthorised access (site staff and public).
- Determining the weight of the load (and lifting tackle used to lift it).
- Locating the load within the load radius capacity of the crane.
- Ensuring the load is free and clear before lifting (remove all fastenings).
- Ensuring the lifting devices are fixed to keep the load level and straight.
- Safety helmets to be worn by all persons in the area.
- Other PPE including rigger gloves, safety footwear and Hi-Vis clothing.
- At all times:
    - Keep persons clear from beneath the load.
    - Ensure good communication between driver, slinger and banksman.
    - One appointed person must be in control of the lift.

Lifting Accessories

- **Ropes and Wires**
    - Rope and wire slings must be in good condition, not kinked, corroded or frayed.
    - Keep ropes and wires off sharp edges of loads by packing.
    - Do not put knots or hitches in slings or lifting ropes.
    - Do not allow them to slip when slack is taken up.
    - Do not drag ropes or wires across the ground.
    - Wear cut-resistant gloves to handle wires and all accessories.
- **Chains**

    Do **not**:
    - Use chains with links locked, stretched or without free movement.
    - Hammer distorted or dislocated links into position.
    - Use corroded, worn, stretched or pitted chains.
    - Cross, twist, kink or knot any chain.
    - Drag a chain from under a load.
- **Connecting Equipment**
    - Do not use if there are signs of damage.
    - Do not use hooks if distorted, stretched or cracked.
    - Ensure safety catches operate and close correctly.
    - Ensure eyebolts are in-line and not distorted.
    - Do not use eyebolts with damaged or distorted threads.
    - Rings and shackles must not be stretched, distorted or cracked.

8.3 Load-Handling Equipment

Hoists

These are static items of equipment for raising and lowering goods (on an open platform) or people (in an enclosed cage):

- Upright hoists are cantilever types used on construction sites, employing a rope or chains passing over a pulley or series of pulleys at the top. Lifting power will be mechanical where loads are to be transported high or weights are great.
- More sophisticated upright hoists use geared drives and steel cables instead of ropes, or rack-and-pinion geared drives. These are more likely to be used for passenger hoists.
- Geared drives are also more likely on inclined materials hoists on construction sites.

The main hazards associated with the use of lifts and hoists are:

- Falls from height (from a landing level, from the platform or with the platform).
- Being struck by moving objects.
- Being struck by falling objects.
- Striking fixed parts of the structure while riding in a hoist.

Controls

There are legislative requirements for the use of lifts and hoists regarding siting, enclosures and safe operation.

Siting

When siting, check the following:

- Must be on solid ground (concrete base for long-term use).
- Free-standing units should be secured to a fixed structure.
- Suitable and sufficient ties should be used - at every other floor.

Enclosures

Requirements for enclosures:

- Access beneath a lift or hoist should be prevented.
- Open hoist platforms require a wider ground level enclosure.
- Where possible, a lift or hoist shaft should be totally enclosed.
- Enclosure gates should be fitted at each landing level.
- Enclosure gates should be interlocked to prevent opening during operation.
- Enclosure structures must prevent access and trapping of persons.
- Person lifts must be controlled from inside the lift cage.

There are legislative requirements for the use of lifts and hoists

Capacity

Maximum design capacity should be marked on the lift or hoist - either number of persons or weight (of goods) in kg or tonnes.

Safety Devices

For safety devices:

- Multiple ropes, hold-back gears and over-run trip systems are required to prevent free-fall and over-running.
- Additional friction brakes should be fitted on passenger lifts.
- Extra cables should be fitted to passenger lifts.

Load-Handling Equipment 8.3

- Landing stage interlocks should prevent gates being opened.
- Failure alarm systems should be fitted.
- A winch brake should stop hoist movement when control is not in operation.

Operation

Lift/hoist operators should be specially trained in safe use of the device they operate:

- The operator should be able to see all landing levels from the operating position.
- A signalling system is required where visibility is impaired.
- People and goods should not travel on the same lift.
- People must never travel on goods-only hoists.
- Extra care should be taken in windy conditions.
- Loads should be properly secured on a goods hoist platform to prevent them opening, tipping or moving.

Inspection, Examination and Maintenance

Note the following:

- Statutory examination is required every six months by a competent person.
- Examination is to be recorded if lift/hoist is used to carry people.
- Equipment must be:
  - removed from use where faults are found; and
  - re-examined before putting back into use.

Forklift Trucks

These vehicles are very versatile and can be used indoors and out, handle a variety of loads and come in many different types, such as:

- Counterbalance truck.
- Pedestrian counterbalance truck.
- Pedestrian pallet stacker.
- Four-directional truck.
- Side-loading truck.
- Rough terrain counterbalance truck.

A variety of attachments increase the forklift truck's versatility, including bale clamps, drum handlers, working platforms, skips, fork extensions and lifting appliances.

Forklift trucks have small wheels and easily become unbalanced when their forks are raised while carrying a load. This could result in the load tipping or the truck overturning.

A gas (LPG) powered industrial counterbalance truck can handle a variety of loads

Other hazards common with forklifts include:

- The need to reverse in many situations, with visibility problems.
- A raised load obscuring the view of the operator.
- Unsuitability of the truck in some environments:
  - Not designed for some loads.
  - Exhaust fumes from diesel trucks in poorly ventilated areas.
  - Counterbalance or reach trucks in rough terrain (construction sites).

8.3 Load-Handling Equipment

- Hydrogen gas from battery-powered trucks, especially when charging.
- Fire and explosion risk from gas (liquefied petroleum gas (LPG)) trucks.
- Manual handling hazards associated with changing batteries and gas cylinders.
- Poorly maintained brakes, steering, tyres and lights.
- Incorrect fitting and use of attachments.

Rough Terrain Counterbalance Lift Trucks

Similar in design to the industrial counterbalance lift truck, these are equipped with larger wheels and pneumatic tyres (as opposed to solid) giving them greater ground clearance and usually a wider wheelbase. They are much more stable on uneven and soft ground.

Hazards include:

- Refuelling:
    - This poses the risk of exposure to hazardous or flammable substances such as diesel, petrol or LPG, with the consequent risk of fire and explosion.
- Fumes from exhaust gases:
    - Forklift trucks with an internal combustion engine can be run on petrol, diesel or LPG. Hazards arise from emission of exhaust fumes in confined spaces or spaces that are inadequately ventilated, causing the risk of asphyxiation.
- Noise:
    - Noise can affect the driver's hearing or distract the driver's attention when the vehicle is in motion.
- Hazards to pedestrians:
    - Being run over, struck by or crushed between the forklift truck and another object such as a wall or load - particularly during reversing.
    - Being struck by an incorrectly secured load falling from the forks.
- Use in adverse weather conditions:
    - Very wet or icy conditions can lead to the truck slipping into trenches and excavations or striking plant or structures due to reduced braking capacity.

Controls

The following safety measures must be adhered to for all forklift truck operations:

- Assess the load and suitability of the forklift.
- Position forks for the load.
- Apply handbrake during lifting and lowering. Do not tilt mast.
- Avoid unstable or uneven ground.
- Remove keys when vehicle is not in use.
- Do not carry passengers (unless seat or work platform is in place).
- Park safely - do not create an obstruction.
- Do not leave trucks unattended on a gradient.
- Closely supervise tandem lifts using two trucks.
- Use a banksman in closely supervised operations.
- Maintain trucks in a safe condition.
- Ensure statutory examinations (**LOLER**) are carried out.

Load-Handling Equipment | 8.3

In addition, the forklift operator should carry out a daily safety check on the vehicle, to ensure that:

- Pneumatic tyres are inflated to the correct pressure.
- Tyres are not damaged or cut and have no foreign objects embedded.
- Parking brake, service brake and steering all work correctly.
- Fuel, water and oil in diesel/petrol engines are at correct levels.
- Batteries in battery trucks are fully charged and leak-free.
- Gas cylinders in LPG trucks have adequate gas, no leaks and cylinders are securely fixed.
- All lifting, tilting and extension systems work correctly.
- Hydraulic systems are charged, have no leaks and hoses are undamaged.
- Audible warnings, flashing beacons and lights all work.
- ROPS and/or restraints are all in safe condition.

Ensure all lifting, tilting and extension systems work correctly

Rough Terrain Counterbalance Lift Trucks

Trained and certified operators and banksmen are critical to safe operation of these vehicles. As well as the safety measures shown above, extra care is needed at the additional heights these vehicles can operate to.

Rough terrain forklifts require:

- Safety devices within the hydraulic system to limit lifting capacity to its safety (weight) rating.
- Height limiting devices.
- A levelling indicator showing danger zones where the load cannot be raised.
- Outriggers and stabilisers (where fitted) with interlocking devices to maintain position in case of failure.
- An indicator lamp to show when outriggers are on firm ground.
- Inspection and examination according to **LOLER**.

Pallet and Sack Trucks

Sack Trucks

Sack trucks are two-wheeled trolleys used for sacks, boxes and small stackable loads.

Pallet Trucks

Flat-version fork trucks:

- With a mechanical fork raising and lowering facility to raise loads from the ground.
- Steered by hand to give good manoeuvrability.

Hazards include:

- Overloading (making them hazardous to move).
- Runaway - they don't usually have brakes.
- Instability of loads (especially on sack trucks).
- Difficult to move over steps and uneven ground due to small wheels.
- Loss of control on or across slopes and ramps.
- Two-wheeled platform trucks may tip and spill a load.
- Careless parking may cause collisions and trips.

8.3 Load-Handling Equipment

For pallet trucks, additional hazards include:

- Potentially not starting in the right direction, due to having pivoted front wheels.
- Tipping if forks are not central beneath a load.
- Trip hazard if not parked properly.
- The forks and the load dropping due to a hydraulic failure.
- Feet becoming trapped beneath the forks or load when lowered.

Controls

Safe use of sack trucks and pallet trucks will require the following measures:

- Plan a safe route avoiding obstructions and slopes.
- Look out for and avoid other pedestrians.
- Put ramps over steps and raised obstacles.
- Ensure the device will carry the weight of the load.
- Place the load centrally and make sure it is secure.
- Use the pallet truck brake when it is stationary, or use chocks.
- Avoid getting feet beneath loads being lowered.
- Stow pallet truck safely with forks out of the way.
- Inspect pallet trucks at regular intervals - record maintenance and tests.
- Operators to use appropriate PPE - in particular, safety footwear, and gloves and aprons to protect while handling loads.

Pallet trucks in a warehouse

Telescopic Handlers (Telehandlers)

Telehandlers are fitted with a pivoted boom on the rear of the truck which allows raising and lowering and telescopic retraction. These machines may be two- or four-wheel drive, and have two- or four-wheel, or crab steering. They are used mainly in agriculture and in the construction industry, with a range of industrial and agricultural attachments available.

Hazards associated with telehandlers include:

- Overturning.
- Impact with people, materials, scaffolding and structures.
- Falling into trenches or holes in the ground.

Controls

Telehandlers require:

- ROPS and driver restraint systems.
- A forward stability indicator with an audible alarm if the load exceeds 96% of the Safe Working Load (SWL).
- Outriggers and stabilisers (where fitted) with interlocking devices to maintain position in case of failure.
- An indicator lamp to show when outriggers are on firm ground.
- Inspection and examination according to **LOLER**.

Telescopic materials handler (telehandler)

Requirements for Lifting Operations

The **Lifting Operations and Lifting Equipment Regulations 1998 (LOLER)** determine the requirements for safe lifting operations. The key requirements are:

> **DEFINITION**
>
> **LIFTING EQUIPMENT**
>
> Work equipment for lifting or lowering loads and its attachments for anchoring, fixing or supporting it, i.e. cranes, goods lifts and hoists, mobile elevating work platforms, vehicle hoists and forklift trucks. 'Lifting accessories' are included, which attach the load to the machine, such as ropes, chains, slings, eyebolts, etc.

- **Strong, Stable and Suitable Equipment (Regulation 4)**

  An employer must ensure that lifting equipment is manufactured from materials suitable for the conditions under which it is to be used.

- **Positioned and Installed Correctly (Regulation 6)**

  This is to reduce the risk of the equipment or a load striking a person, or a load drifting, falling freely or being released unintentionally.

- **Visibly Marked with Safe Working Load (Regulation 7)**

  Machinery and equipment must indicate a Safe Working Load (SWL). Where different equipment configurations affect this, each configuration must be indicated. Accessories are also to be marked with SWL.

- **Plan and Supervise (Regulation 8)**

  An employer must organise, plan and supervise lifting operations and ensure they are carried out by competent persons in a safe manner. Those planning operations are to be suitably competent.

- **Equipment for Lifting People (Regulation 5)**

  Equipment must be strong, stable and suitable. People must not be struck, trapped or fall when they are lifted. Doors, cages and alarm systems for summoning assistance are to be safe and in place.

Periodic Inspection and Examination/Testing of Lifting Equipment

LOLER covers the requirement for most items of lifting equipment and accessories to undergo regular visual inspection and thorough examination by a competent person.

Regular Visual Inspection

Regular visual inspection should be conducted before use and weekly.

An examination scheme should determine where defects are likely to be detected, including:

- Rapid wear caused by use in severe environments.
- Wear through repeated operations.
- When an item may malfunction.
- Where tampering with a safety device may occur (e.g. an interlock).

8.3 Load-Handling Equipment

Thorough Examination

Thorough examination is a requirement of **LOLER**, Regulation 9. It should be carried out:

- When lifting equipment is used for the first time or for the first time at a new location - to ensure it is correctly installed and free from defects.
- Every 6 months - for all lifting equipment and accessories for lifting people.
- Every 12 months - for all other types of lifting equipment.
- Following any incident or accident that might have stressed the equipment.
- Following any change in conditions of use which could affect the safety of the equipment.

Ropes and chains on equipment used to lift people are to be inspected every working day (**LOLER** Regulation 5).

Records

The following checks should be made:

- A written examination scheme must be in place scheduling these inspections and examinations.
- Where a defect is found that could cause injury, Regulation 10 requires a report to be made to the employer, who must not allow the equipment to be used until the defect is rectified.
- Defects causing imminent or serious danger are also reported to the HSE.
- Reports of each inspection must be kept until completion of the next inspection and receipt of its report.

STUDY QUESTIONS

9. What are the most common hazards associated with:
   (a) Forklift trucks?
   (b) Hoists?
   (c) Cranes?
   (d) Sack trucks?

10. What personal protective equipment might be appropriate for working with:
    (a) Pallet trucks?
    (b) Cranes?

11. When, under the **Lifting Operations and Lifting Equipment Regulations 1998**, must a thorough examination of lifting equipment used to carry people be conducted?

(Suggested Answers are at the end.)

Summary

This element has dealt with some of the hazards and controls relevant to manual and mechanical handling.

In particular, this element has:

- Explained the meaning of musculoskeletal disorders and work-related upper limb disorders and shown examples of repetitive construction activities that can cause MSDs and WRULDs.
- Discussed the ill-health effects of poorly designed tasks and workstations, together with the factors contributing to ill-health conditions including the task, environment and equipment, and appropriate controls such as matching the workplace to the needs of the worker.
- Explained the hazards and control measures which should be considered when assessing risks from manual handling activities, including:
    - identifying common manual handling injuries;
    - assessing the task, individual, load and environment (TILE); and
    - looking at efficient movement principles.
- Explained the hazards and control measures to reduce the risks in the use of mechanical and hand-operated lifting and moving equipment, the requirements for lifting operations to be conducted safely, and requirements for regular visual inspection and statutory examination of lifting equipment.

Exam Skills

Question

Scenario

You have received complaints from bricklayers about musculoskeletal disorders (MSDs). You decide to conduct a safety tour of the area where bricklayers are working and you notice work practices that could exacerbate and contribute to MSDs. You speak to some of the workers about their work technique and they are unaware of how MSDs can occur and how to ensure they reduce the risk of developing injuries.

Task: Musculosletal Health and Load Handling

Back in the office you decide to prepare a briefing document to be used at a toolbox talk on the subject of MSDs. In the briefing document you need to create two clear headings:

(a) Why bricklayers may be at risk of MSDs. **(6 marks)**

(b) What control measures can be introduced to help reduce the risks to bricklayers. **(4 marks)**

(Total: 10 marks)

Approaching the Question

Now think about the steps you would take to answer this question:

Step 1 The first step is to **read the scenario carefully**. Note that complaints have been made by the bricklayers and obviously they are unclear how their work practices can cause harm but also how to improve their working technique.

You decide to create a briefing document for a toolbox talk so you will need to structure your notes using the two headings given.

Step 2 Now look at the **task** - prepare some notes under the two headings.

Step 3 Next, consider the **marks** available. In this task, there are 6 marks available for the first part and 4 marks for the second part of the question. Tasks that are multi-part are often easier to answer because there are additional signposts in the question to keep you on track. In this task, you have to create a briefing document that is easy to understand; giving examples for each part can aid understanding. You will need to provide around 8 or 9 different pieces of information including examples for this task. In this answer, some pieces of information may gain more than 1 mark as they will require additional detail. The headings will allow you to keep your response separate – this will also help the examiner when marking.

Step 4 **Read the scenario and task again** to make sure you understand the requirements and ensure you have a clear understanding of musculoskeletal disorders. (Re-read your study text if you need to.)

Step 5 The next stage is to **develop a plan** - there are various ways to do this. Remind yourself, first of all, that you need to be thinking about the causes of MSDs and the measures that can reduce the effects of them.

Exam Skills

Suggested Answer Outline

Factors that might contribute to MSDs in bricklayers:

- Repetition of the task.
- Weight of the load (small).
- Twisting and turning action.
- Posture (bending and laying at ground level working up to stretching).
- Climatic conditions (weather cold and rain).
- Current medical conditions.

Measures to reduce the risk of developing MSDs:

- Layout of the workstation (position of brick pile in relation to the work).
- Wearing of PPE (gloves, safety boots, warm weatherproof clothing).
- Regular breaks with physical preparation of the body.

Now have a go at the question yourself.

Example of How the Question Could be Answered

(a) *The key points to cover in a briefing document for a toolbox talk on how work-related upper limb disorders can occur whilst undergoing bricklaying activities are firstly how repetitive the task is. Bricklaying involves picking up bricks which are generally light in weight and compact to hold. The risk is when during the task you are having to bend and twist due to the position of the brick stockpile and where the brick is being placed. This action will exacerbate a back condition due to the frequency of repetition. The risk will be increased if the bricklayer is bending down or having to reach in order to do the work.*

As the work is often conducted in an outdoor setting the elements such as wind, rain and low temperature can cause the body to cool and in doing so lead to possible muscle strains and sprains due to the circulatory system reducing blood flow rate to the extremities of the body.

Finally, the overall medical condition of the bricklayer needs to be considered as if he has sustained previous musculoskeletal injuries, these can be exacerbated by this type of activity.

(b) *Control measures that can be introduced to reduce the risk of MSDs while conducting bricklaying activities would include layout of the workstation where the pile of bricks is placed at a height similar to that of where the bricks are being laid. This will help in reducing the twisting, bending and stretching action as well as improving posture, especially if the bricklayer can be in a standing position.*

Ensuring the bricklayer is wearing appropriate clothing (gloves, warm and waterproof clothing) especially when working in wet and cold conditions. This will insulate the body and ensure body extremities are kept warm maintaining circulation which will reduce sprains and strains.

As brick laying is physical regular breaks should be planned with appropriate facilities provided where they are able to sit down in a warm environment. Before commencing work, muscle preparation needs to take place through stretching and working the muscles through exercises should take place.

Reasons for Poor Marks Achieved by Exam Candidates

- Not following a structured approach for the briefing document: failing to provide information on the two subject areas.
- Not expanding the answer beyond a few words as opposed to giving a sentence of explanation.
- Misinterpreting the question by focusing on work-related upper limb disorders and control measures, even though very similar.

Element 9

Work Equipment

Learning Objectives

Once you've studied this element, you should be able to:

1. Outline general requirements for work equipment.

2. Outline the hazards and control measures for hand-held tools, both powered and non-powered.

3. Describe the main mechanical and non-mechanical hazards of machinery.

4. Explain the main control measures for reducing risk from common construction machinery hazards.

5. Describe the hazards of working near water.

Contents

General Requirements for Work Equipment	**9-3**
Introduction to the General Requirements for Work Equipment	9-3
Scope of Work Equipment	9-3
Providing Suitable Work Equipment	9-4
Preventing Access to Dangerous Parts of Machinery	9-5
When the Use and Maintenance of Equipment with Specific Risk Needs to be Restricted	9-16
Providing Information, Instruction and Training about Specific Risks	9-16
Why Equipment Should be Maintained and Maintenance Conducted Safely	9-17
Emergency Operations Controls	9-19
Hand-Held Tools	**9-22**
Considerations for Selecting Hand-Held Tools	9-22
Hazards of a Range of Hand-Held Tools	9-23
Machinery Hazards and Control Measures	**9-26**
Consequences as a Result of Contact with Hazards Identified in ISO 12100:2010	9-26
Hazards and Controls of a Range of Site Equipment	9-30
Working Near Water	**9-35**
Additional Appropriate Control Measures	9-35
Summary	**9-39**
Exam Skills	**9-40**

General Requirements for Work Equipment

IN THIS SECTION...

- The general requirements for work equipment include:
  - Understanding the scope of work equipment and its suitability, including UK conformity requirements.
  - Restricting the use and maintenance of equipment which presents specific risks, and the provision of information, instruction and training for people at risk.
  - Ensuring safe maintenance of equipment.
  - Ensuring regular visual inspection and statutory examination of work equipment are carried out and appropriate records kept.
  - Appreciating the importance of emergency controls; stability of equipment; lighting, markings and warnings for safety purposes; and unobstructed work space.

Introduction to the General Requirements for Work Equipment

The key statutory requirements in respect of this area of health and safety are contained in the **Provision and Use of Work Equipment Regulations 1998 (PUWER)**. The **Supply of Machinery (Safety) Regulations 2008** are also relevant.

Scope of Work Equipment

DEFINITIONS

WORK EQUIPMENT

"*...any machinery, appliance, apparatus, tool or installation for use at work (whether exclusively or not).*"

PUWER 1998

This includes an assembly arranged and controlled to function as a whole (e.g. bottling plant); equipment provided by an employer; **and** tools brought in by employees (but excludes privately owned vehicles).

USE

"*...any activity involving work equipment...includes starting, stopping, programming, setting, transporting, repairing, modifying, maintaining, servicing and cleaning.*"

PUWER 1998

Work equipment will include:

- **Single machines** - e.g. drilling machines, woodworking machines, abrasive wheels, photocopiers, dumper trucks and circular saws.
- **Hand tools** - e.g. hammers (claw, slater, club, brick), screwdrivers (straight, Phillips, pozidrive), files, chisels, picks, shovels, spanners, wrenches, saws (hand, tenon, hacksaws).
- **Power tools** - which can be powered by compressed air, electricity, explosive cartridge or the internal combustion engine, e.g. concrete breakers, rock drills, air grinders, disc cutters, sanders, nail guns and portable electric drills.
- **Vehicles** - where they are used within workplaces (i.e. off public roads), therefore including lift trucks, cranes, excavators, etc.

9.1 General Requirements for Work Equipment

Providing Suitable Work Equipment

The employer must ensure that work equipment:

- Is appropriate for the work to be undertaken.
- Is used in accordance with the manufacturer's specifications and instructions.
- If adapted, is still suitable for its intended purpose.

The location in which the work equipment is used must be assessed to take into account any risks from particular circumstances, e.g. electrically powered equipment used in wet or flammable atmospheres.

The equipment must be designed ergonomically, to take into consideration its compatibility with human dimensions.

Conformity with Relevant Standards

All new machinery must be designed and constructed to meet minimum safety standards.

Under the **Supply of Machinery (Safety) Regulations 2008**, a manufacturer may not supply machinery for use **unless it is safe** and, in particular:

- It satisfies certain essential health and safety requirements.
- A technical file is compiled (this is like a portfolio of evidence).
- Information is provided on its safe operation, e.g. instructions.
- The appropriate conformity assessment procedure has been followed.
- The manufacturer has issued a Declaration of Conformity (i.e. stating that the product conforms to all relevant requirements and referencing any relevant standards used in that assessment).
- The conformity assessment mark has been properly affixed to the machinery.

Businesses wishing to sell goods in UK markets must meet the UK rules of product conformity and, in most cases, from 1 January 2025 show this by a United Kingdom Conformity Assessed (UKCA) or United Kingdom Northern Ireland (UKNI) marking. Manufacturers may only affix UKCA or UKNI markings when all the requirements of all the UK product supply legislation applicable to the product have been met.

The UKCA (UKNI) marking system is similar in most respects to the European Union CE marking.

If a machine is built fully in accordance with one or more published harmonised standards (i.e. EN standards that have had their reference published in the official EU journal), there is a presumption of conformity with essential health and safety requirements.

CE mark

UKCA mark

General Requirements for Work Equipment | 9.1

In applying the UKCA (UKNI) mark to the product and signing the Declaration of Conformity, the manufacturer is declaring that the product meets the requirements of all the standards which apply to it. This provides some assurance that it is safe when properly installed, maintained and used for its intended purpose.

The UKCA (UKNI) mark must be affixed to (in order of preference) the product, its instruction manual or its packaging.

Note: until 31 December 2024, businesses wishing to sell goods in UK markets may use either the UKCA/UKNI mark or the CE mark.

In many cases, the assessment of machinery safety is done entirely by the manufacturer, without any independent third-party involvement. In these cases it is self-assessment and therefore just a claim to safety.

Manufacturers must also ensure that their construction product bears a type, batch or serial number that positively identifies it, and (depending on the size and nature of the product) that this information is provided on the packaging or in a separate document.

PUWER places a duty on **users** of work equipment to check that it is safe before use. In practice, this means that the employer wishing to use the equipment should:

- Undertake a visual check of the machine (guarding, etc.).
- Confirm that:
  - The machine is UKCA/UKNI or CE -marked.
  - There is a Declaration of Conformity.
- Ensure that:
  - Operating instructions have been provided.
  - There is information about residual hazards such as noise and vibration.

UKNI mark

Preventing Access to Dangerous Parts of Machinery

'L22 - *Safe use of work equipment*', the ACoP which supports the **Provision and Use of Work Equipment Regulations (PUWER)**, in regard to Regulation 11 says that effective measures must be taken to prevent access to dangerous parts of machinery or stop their movement before anyone enters a danger zone.

Main Types of Safeguarding Devices

This section will identify the characteristics, key features, limitations and typical applications of a range of guards.

Fixed Guards

Enclosing

These are guards with no moving parts, and are designed to prevent access by enclosing the hazard. There may be access points where materials can be inserted and withdrawn, and hatches for maintenance or inspection.

9.1 General Requirements for Work Equipment

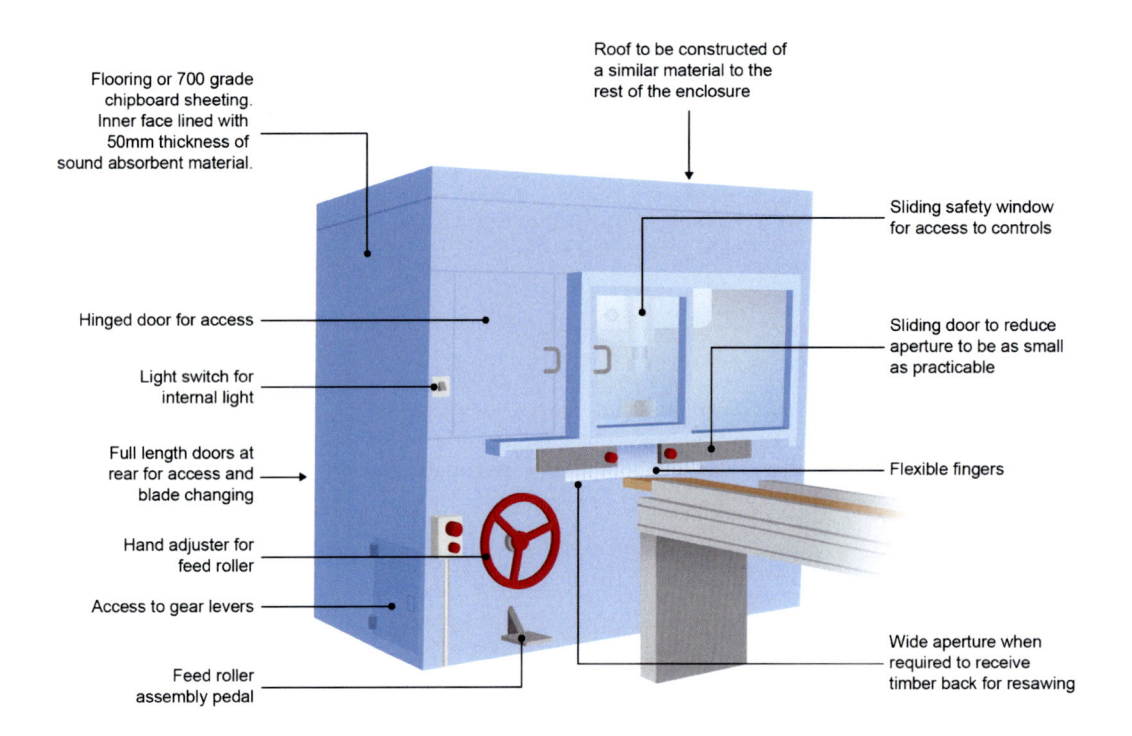

Fixed enclosed guard

Adjustable

This type of guard has no moving parts and is to keep the person away from the hazard. The false table shown in the following diagram serves as a fixed distance guard.

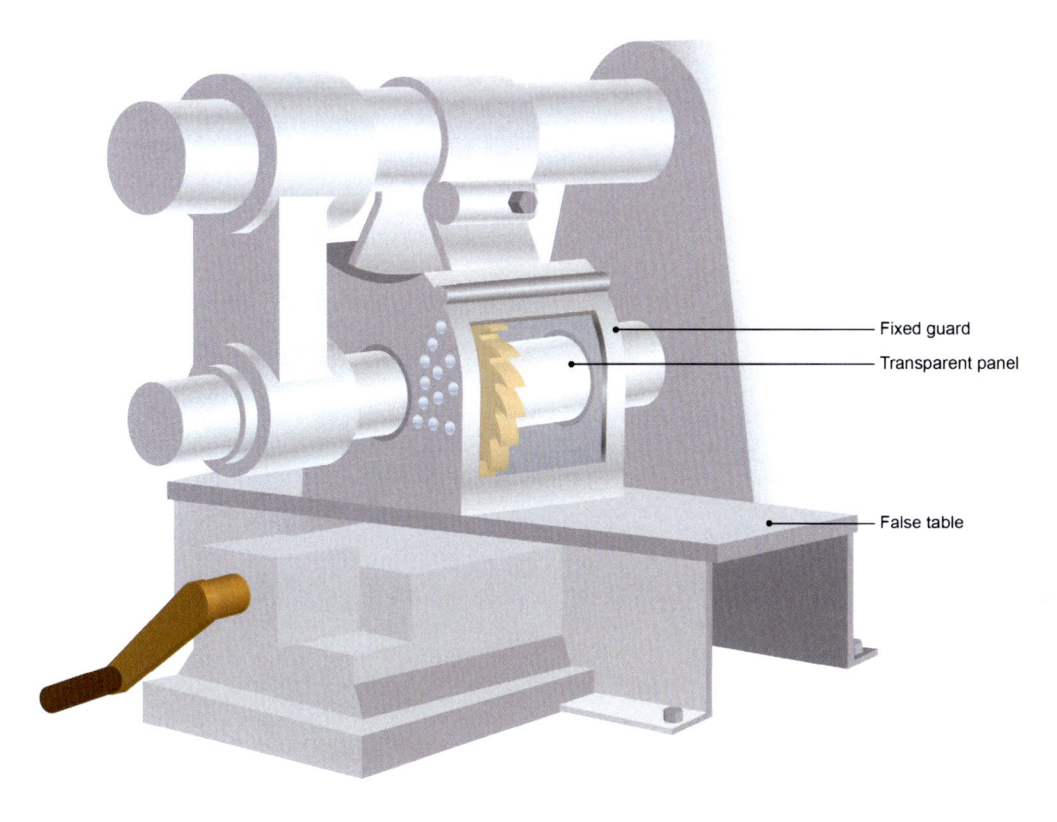

Fixed distance guard

A fixed distance guard is a fixed guard that does not completely enclose a hazard, but which reduces access by virtue of its physical dimensions and its distance from the hazard.

Another type of distance guard which completely surrounds machinery is commonly called a perimeter-fence type guard. An example of this is the enclosure around a robotic area.

When it is necessary for work to be fed through the guard, openings should be sufficient only to allow the passage of material and should not create a trap between the material and the guard. If access to the dangerous parts cannot be prevented by the use of a fixed enclosed guard with a plain opening, then a tunnel guard of sufficient length should be provided.

Interlocked Guards

When an operator is required to enter a hazardous area of machinery and where fixed guards are not a practical option, interlocked guards are the next best protective device. Interlocked guards are defined as a guarding system which, when the hazard area is open, prevents the machinery from operating. Implicit in this definition are three important points which control the design and operation of an interlocked guard:

- It must prevent movement of the dangerous parts of the machine when the hazard area is open.
- It must not allow access to the hazardous area until the potential hazard has been made safe.
- It must not allow the machinery to operate until the guarding system is fully operational.

Other factors of importance are:

- If the interlocked system should fail, it should fail in such a way that the system remains safe.
- The interlocked system should be difficult to defeat.

9.1 General Requirements for Work Equipment

The operation of an interlock may be electrical, mechanical, hydraulic or pneumatic. The choice is often dependent on the power medium (e.g. hydraulic) in use to operate the machine. In more complex machines, a combination of interlocks may be in place.

> **TOPIC FOCUS**
>
> An example of the use of an interlocked guard is the enclosure of a CNC lathe or grinder. This guard fits around the workspace within which the machine operates, and performs several functions:
>
> - It keeps the operator out of the danger area of the machine during normal operation and prevents contact with dangerous moving parts. Specifically, it prevents entanglement with rotating parts and contact with sharp parts that present a significant cutting/severing hazard.
> - It protects the operator from ejected parts such as swarf or fine particles of grinding wheel.
>
> It also offers some protection against the non-mechanical hazards, such as:
>
> - Noise.
> - Aerosols - such as cutting fluids used to cool and lubricate the wheel.
>
> Operators of CNC machines have to enter the machine enclosure routinely during normal operations and therefore it would be inappropriate to rely on a fixed guard for safety. Entry into the enclosure is required for:
>
> - Cleaning - to remove swarf, dust and other contaminants which may affect normal machine operation if left in the enclosure.
> - Setting - configuring the machine into the right position in order for it to start.
> - Changing tools - CNC machines frequently use multiple tools during routine work.
> - Changing the workpiece - at the start and end of each work cycle.
>
> Routine inspection of the workpiece and tools is important to check quality and to ensure no problems are occurring.

Automatic Guards

> **HINTS AND TIPS**
>
> You should note carefully the description of an automatic guard, as many candidates tend to confuse 'automatic guards' and 'interlock guards in examination' when describing different guards types.

Automatic guards may be defined as 'guards which **forcibly move** persons from the hazard area (sweep away) **before** the machinery operates'. In theory, the person should not be able to enter the hazard area while the automatic guard is operating.

General Requirements for Work Equipment | 9.1

As the guarding system uses motion as an essential part of its protective mechanism, doubts as to the acceptability of such a system must be raised. Apart from the concept of motion being the fundamental cause of machinery hazards, there are practical considerations, such as:

- The speed at which the guard has to operate to overtake the hazard may be dangerous.
- Tall persons may fall or lean over the guard into the hazard area.

A typical arrangement for an automatic guard on a power press is shown in the next diagram.

9.1 General Requirements for Work Equipment

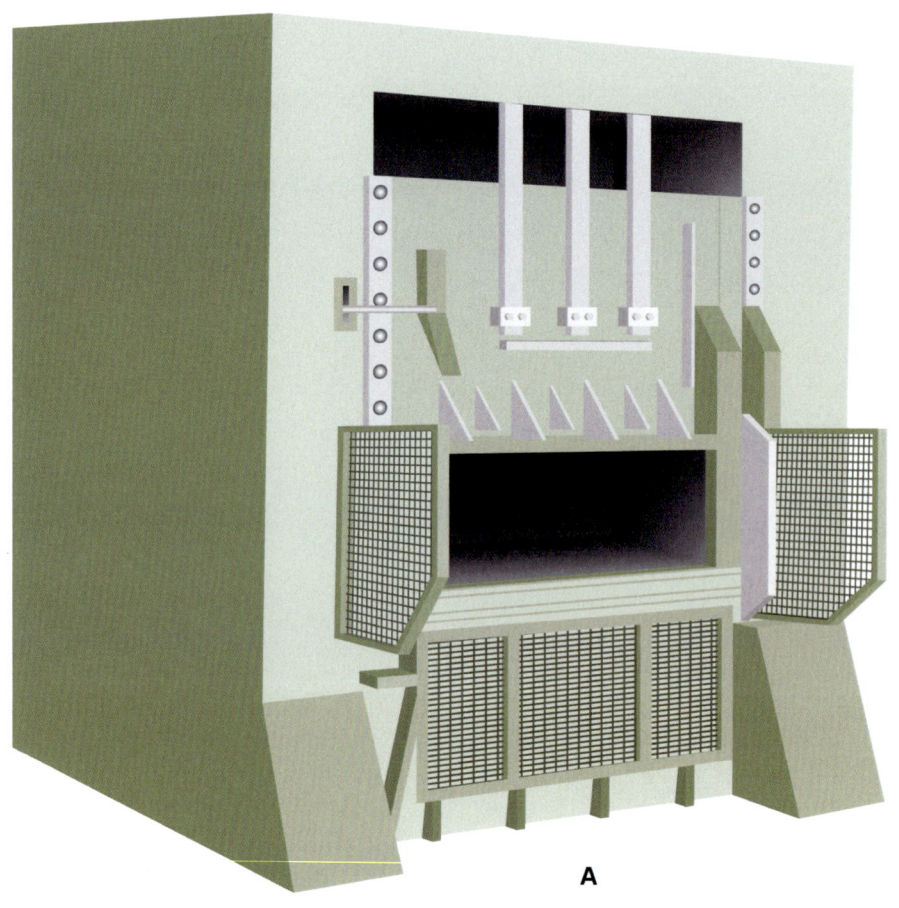

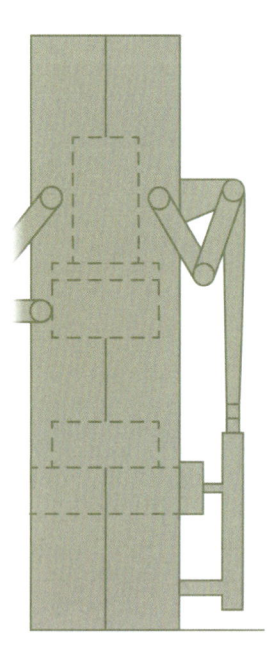

A

B Guard "in" position

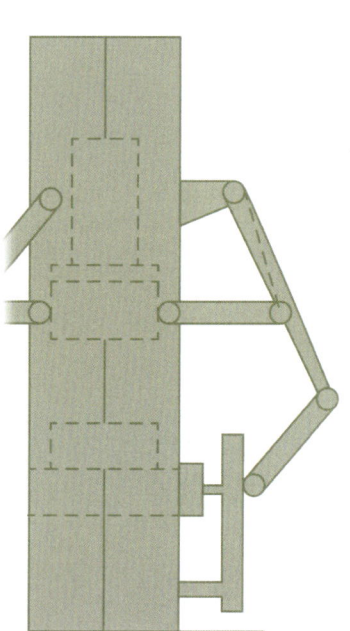

C Guard "out" position

Automatic guard

Trip Devices

Safety trip controls provide a quick means for deactivating the machine in an emergency situation. A pressure-sensitive body bar, when depressed, will deactivate the machine. If the operator or anyone trips, loses balance, or is drawn toward the machine, applying pressure to the bar will stop the operation. The positioning of the bar, therefore, is critical. It must stop the machine before a part of the employee's body reaches the danger area.

Trip Bar for Radial or Pillar Drills

The next diagram illustrates a trip bar guard which can be fitted to pillar drills. A micro-switch attached to the trip bar will, if slightly displaced, cut off AC supply and inject DC into the motor, so that it stops instantly.

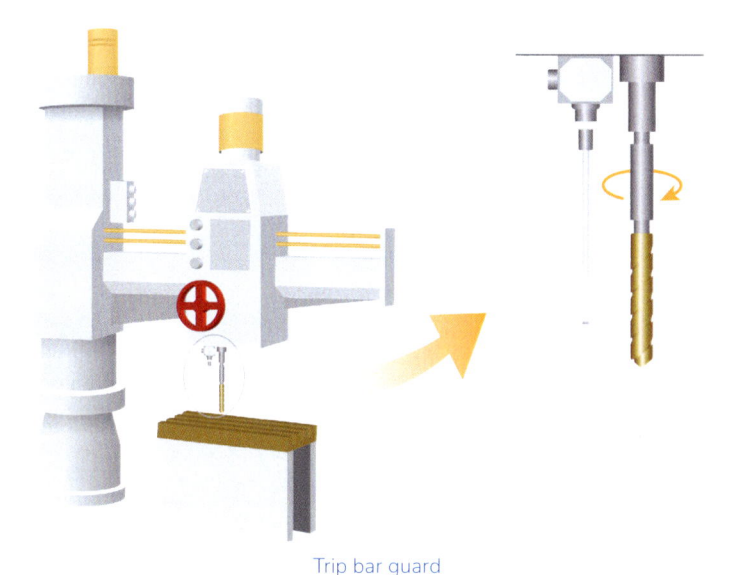

Trip bar guard

With this type of system, employers should ensure:

- Maintenance of maximum sensitivity for the trip bar, i.e. the minimum of movement is required to activate the micro-switch.
- Monitoring of the micro-switch for contact wear.
- That the trip guard is not being used as an **operational** brake for the drill.

9.1 General Requirements for Work Equipment

Photo-Electric Guards

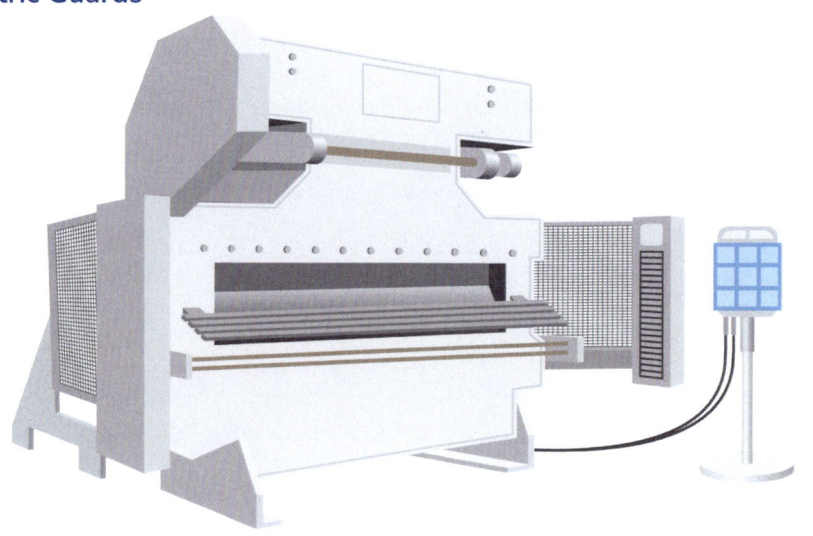

A photo-electric guard fitted to a press brake. Based on original source L22 Safe use of work equipment (4th ed.), HSE, 2014 (www.hse.gov.uk/pubns/books/l22.htm)

Another form of trip device used for press brakes and hydraulic presses is the photo-electric guard.

The photo-electric (optical) presence-sensing device uses a system of light sources and controls which can interrupt the machine's operating cycle. If the light field is broken, the machine stops and will not cycle. This device must be used only on machines which can be stopped before the worker can reach the danger area. The design and placement of the guard depends on the time it takes to stop the mechanism and the speed at which the employee's hand can reach across the distance from the guard to the danger zone.

Cam-Activated Switches

A cam-activated switch usually works by the movement of a raised part of the switch arrangement against a spring-loaded push rod. The push rod will normally have a roller at the end of it.

As the raised portion of the switch comes into contact with the roller, the push rod is depressed against the spring pressure. Depending on the operation of the switch, this will either make or break the contact, or open or close the valve. As the raised portion moves away from the roller, the spring reasserts itself.

A typical example of the arrangement and operation of linear cam-activated switches is shown in the electrical limit switch diagrams which were seen earlier.

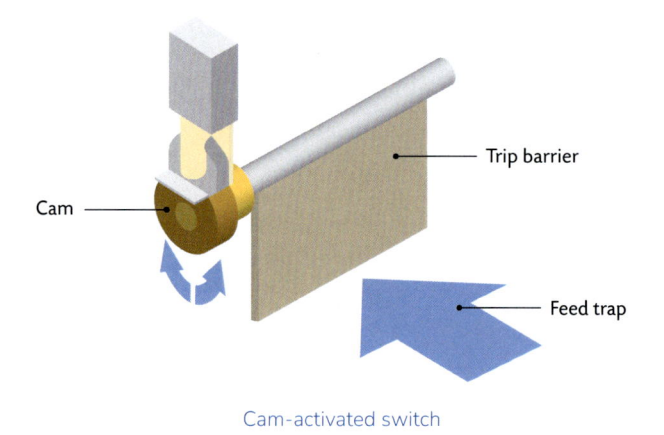

Cam-activated switch

Adjustable Guards

Adjustable guards are guarding systems which require manual adjustment to give protection. They are used on woodworking machinery, milling machines, lathes, drills and grinders. Where it is impracticable to prevent access to the dangerous parts because they are unavoidably exposed during use, e.g. the cutters on milling machines and the cutters of some woodworking machines, the use of an adjustable guard may be permissible in conjunction with other closely supervised conditions, e.g. a sound floor, good lighting and adequate training of the operator.

General Requirements for Work Equipment | 9.1

An adjustable guard provides an opening to the machinery through which material can be fed, the whole guard or part of it being capable of adjustment in order that the opening can be varied in height and width to suit the dimension of the work in hand. It is essential in such cases that the adjustment is carefully carried out by a suitably trained person. Regular maintenance of the fixing arrangements is necessary to ensure that the adjustable element of the guard remains firmly in place once positioned. The guard should be so designed that the adjustable parts cannot easily become detached and mislaid.

Many of the guards are designed so the workpiece can be observed during the machine operation. Windows of perspex, polycarbonate or armoured plate glass allow the operator a clear view. Some systems are made with a telescopic fencing or a slotted movable casting, both systems allowing observation of the workpiece.

The following adjustable guarding systems are in common use:

- **Circular Saws**

    The cover is adjusted so that the height is large enough for the workpiece to be cut by the saw.

- **Vertical Drills or Woodworking Moulder**

    The transparent cover which allows a clear view of the drill or cutter is adjusted and secured in position by the thumb screw.

- **Abrasive Wheels**

    A simpler version of an adjustable guard can be used on abrasive wheels.

The use of adjustable guards is allowed in many situations by the enforcing authorities but critical thought should be given before they are used as a guarding system. Their main weakness lies in the fact that they are controlled by the machine operator, and not by the person or organisation responsible, by law, for controlling the safety of the workplace. As a consequence, there are two potentially serious risks:

- They can easily be defeated.
- They rely on operators being 100% vigilant in providing for their own safety - a condition the guard should provide, not the operator.

Where adjustable guards are used, strict training and supervision of operators is of paramount importance.

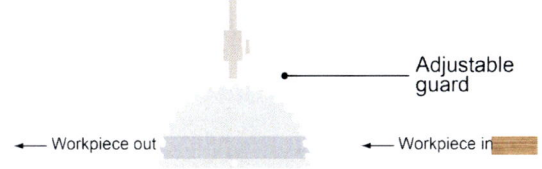

Circular saw with adjustable guard

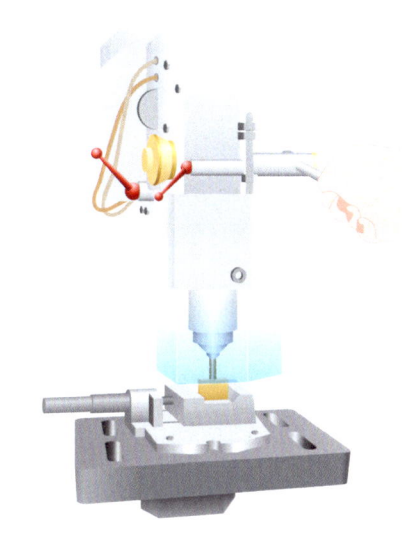

Drilling machine chuck guard

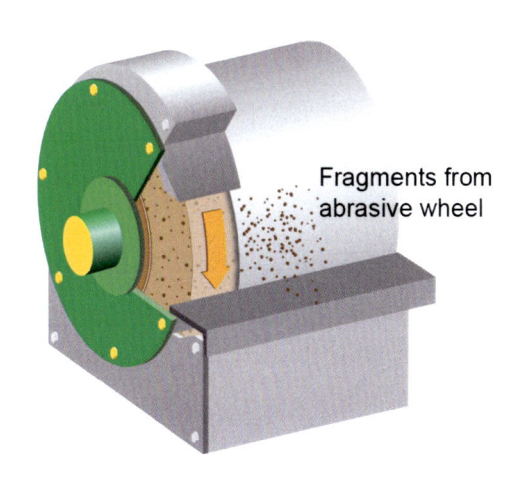

Abrasive wheel guard

Self-Adjusting Guards

A self-adjusting guard is a fixed or movable guard which, either in whole or in part, adjusts itself to accommodate the passage of material, etc.

This type of protection is designed to prevent access to the dangerous part(s) until actuated by the movement of the workpiece, i.e. it is opened by the passage of the workpiece at the beginning of the operation and returns to the safe position on completion of the operation.

Consideration should be given to the use of feeding and take-off devices, jigs and fixtures when this type of guard is used.

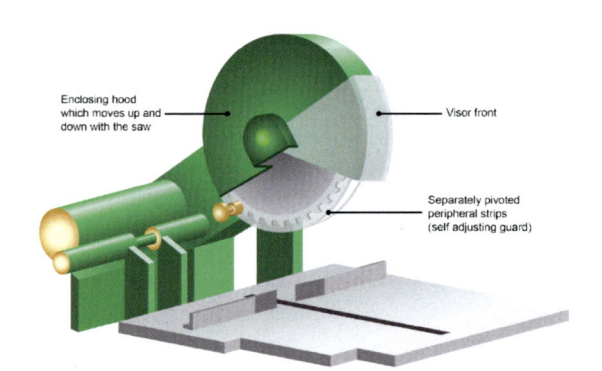

The diagram shows a typical arrangement. The guard rests on top of the work and closes fully when removed. Note that this kind of guarding can be difficult for the operator and is easy to defeat. However, it is sometimes the only practicable method.

Typical arrangement of a self-adjusting guard

Two-Hand Controls

Where guarding is impracticable, as in the following diagram, two-hand control offers a means of protecting the hands of the machine operator. It may also be used as a hold-to-run control.

A two-hand control device requires both hands to operate the machinery controls, thus affording a measure of protection from danger to the machinery operator only. It should be designed in accordance with the following:

- The hand controls should be so placed, separated and protected, as to prevent spanning with one hand only, being operated with one hand and another part of the body, or being readily bridged.

- It should not be possible to set the dangerous parts in motion unless the controls are operated within approximately 0.5 seconds of each other.

- Movement of the dangerous parts should be brought to a stop immediately or, where appropriate, stopped and reversed if one or both controls are released while there is still danger from the movement of those parts.

- The hand controls should be situated at such a distance from the danger point that, on releasing the controls, it is not possible for the operator to reach the danger point before the motion of the dangerous parts has been stopped.

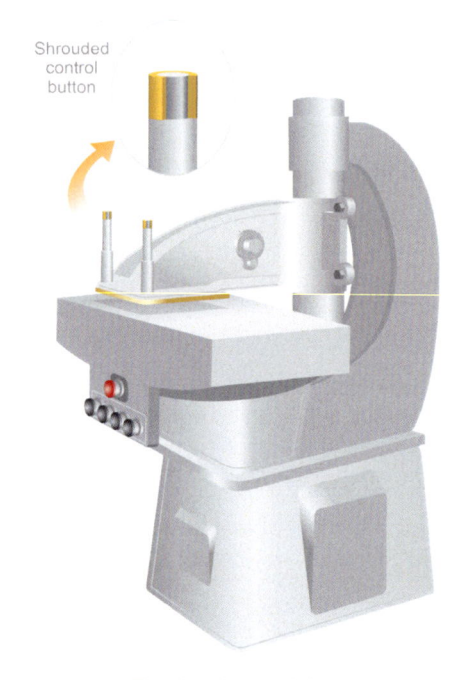

Two-hand control device

There are, unfortunately, difficulties with the use of two-handed controls:

- Experience has shown that most two-handed systems will eventually be defeated by the determined operator.

- During operation, the system protects the operator; third parties are always at risk.

- Often, frequent maintenance is required, as most systems require a complex mechanism in order to make them effective.

The use of two-handed control systems as a method of guarding machinery must be considered as having very limited practical value.

General Requirements for Work Equipment | 9.1

Jigs and Push Sticks

These are designed to hold a workpiece in place when in the machine.

- **Jigs** remove the need for the person to be in the danger zone during operation. They are not, in themselves, a guard, but can be considered as part of the safety measures when assessing the type of guard that is needed on any appropriate machine.
- A **push stick** performs a similar function. A good example is feeding a length of wood into a circular saw. As the hands that are holding the wood approach the saw blade, a push-stick can be used to distance the operator's hand from the moving blade.

Machinery Guarding Guide

The following table provides a summary of the characteristics of different types of machinery guarding.

As machinery becomes more complex and production lines more automated, perhaps involving robots, the demands on the designer become greater and an overall systems approach is essential.

Type	Description	Strengths	Weaknesses	Means of Overriding
Fixed guard	Permanently in place after installation (e.g. welded, riveted). Normally requires a special tool to remove.	Presents the most desirable barrier between operator and hazard. No moving parts. Cannot be interfered with by operator. Virtually maintenance-free.	Machine will still operate with guard removed. Size of holes for material feed may limit operability.	Special tool for removal may be too widely available.
Interlocked guards				
General	System where the guard is integrated with the control system.	Less dependent for their effectiveness on the control of human behaviour. Less easy to defeat. Do not represent the last line of defence as, e.g. a trip device does.	Design of the interlock critical; should be designed to fail to safety.	

9.1 General Requirements for Work Equipment

Type	Description	Strengths	Weaknesses	Means of Overriding
Automatic guard	Moves into position as part of machine cycle.	Pushes any part of a person away from danger area.	Slows operation. Bad design may cause trapping hazard or impact injury.	Mechanical action can be overridden.
Trip devices	Causes machine to stop or become safe when person approaches.	Useful when approach by person required as part of job.	Trips may not be set to cover all means of access. Time delay in stopping machine may be greater than time from tripping guard to reaching machine.	Person can avoid devices as part of 'I can beat this machine' syndrome.
Adjustable/self-adjusting guards	Guard which can be varied in size to suit situation.	Allows variable sized work-pieces.	Easily defeated.	Adjust out of range.
Two-handed control	Two controls activated and released simultaneously.	Keeps operator's hands away from moving parts. Rapid manual movement of guard into place.	Protects only operator's hands, not other parts of body or other people.	Two people, each holding one handle.

When the Use and Maintenance of Equipment with Specific Risk Needs to be Restricted

PUWER Regulation 7 deals with the issue of specific risk. Where risks cannot be controlled adequately by guards or safety devices, use of the equipment should be restricted to those people who have received sufficient information, instruction and training.

Similarly, where risks arise out of maintenance, modification and repair operations, only competent people should be allowed to carry out those tasks.

Providing Information, Instruction and Training about Specific Risks

Where the use of work equipment involves a specific risk to health and safety:

- The use of that equipment must be restricted to persons given the task of using it.
- Any repairs, modifications, maintenance or servicing carried out on that equipment must be undertaken only by competent and trained people who have been specifically designated to perform the work.
- Care must be taken regarding the use of that equipment in certain environments, e.g. confined spaces.

General Requirements for Work Equipment | 9.1

All users of work equipment must be provided with adequate health and safety information, either in writing or verbally, including information regarding specific risks associated with the use of that equipment.

Written information:

- Relates primarily to manufacturers' manuals, warning labels, information sheets, etc.
- Will normally only be required for more complex machinery.
- Must be easily accessible.
- Must be in a form which employees will be able to readily understand.

All persons who use work equipment should receive training on the equipment they are expected to use and be aware of the health and safety implications, potential risks and precautions:

- Young persons (those under the age of 18) should be given special consideration due to their inexperience and immaturity, and be closely supervised by a competent person.
- Managers, supervisors and maintenance staff must also receive adequate training in relation to health and safety, the use of the equipment and risks entailed in its use.

The more complex the equipment used and the bigger the risks, the greater the need for comprehensive information, instruction and training.

Why Equipment Should be Maintained and Maintenance Conducted Safely

Work equipment must be maintained:

- In an efficient state.
- In efficient working order and in good repair, i.e. where a defect, damage or wear is detected, appropriate remedial action must be taken.

Where any machinery has a maintenance log, the log must be kept up to date.

Routine checks and maintenance must be carried out to ensure that all equipment remains safe to use at all times. Maintenance work should be:

- Undertaken only by competent people.
- Carried out without exposing maintenance workers to risks to their health and safety.

Normally, maintenance work is carried out when equipment is shut down. Maintenance on live equipment should be avoided - any power supply, such as electricity, hydraulic or pneumatic power, should be isolated; any blades, teeth, or rotary parts should be carefully handled; replacements must be made with the correct type. PPE may be required if coming into contact with dusty or hazardous materials. Any guards that have been removed for access must be replaced securely. If work cannot be carried out with the machine isolated, a permit-to-work system may be required.

Scheduling maintenance may involve:

- Planned preventive maintenance (monitored; time or running hours based).
- Condition-based maintenance (safety-critical parts).
- Breakdown maintenance.
- Opportunity maintenance - using quiet 'downtime' to maintain.

Employers should keep records of such maintenance and the remedial action taken (there is no general requirement to keep maintenance logbooks).

Regulation 22 of **PUWER** says that steps must be taken to ensure that maintenance operations can be carried out safely. In most cases, maintenance of equipment will have been considered at the design stage but, if not, a risk assessment should be carried out.

9.1 General Requirements for Work Equipment

> **TOPIC FOCUS**
>
> **Planned preventive maintenance** is required where:
>
> - The safety of an item of work equipment depends on the installation conditions.
> - An item of work equipment is exposed to conditions that could cause dangerous deterioration.
>
> The programme should be based on regular inspection of the item and should be recorded.

Requirements for Examination and Inspection

Inspection is required under a number of different circumstances, including:

- **After installation or re-installation** - before being used for the first time or after refitting, to prevent faults from incorrect installation.
- **Where deterioration leads to a significant risk** - e.g. items of equipment left out in all weathers or in a harsh environment (perhaps exposed to chemicals).
- **Where exceptional circumstances may jeopardise safety** - e.g. after major modifications or repair, known or suspected damage, change of use or after long periods of disuse.

Records of such inspections must be kept, either handwritten or electronically.

The level of inspection for general work equipment is less detailed and less intrusive than the type of thorough inspection required for power presses (**PUWER**, Regulation 33) or some types of lifting equipment (**LOLER**, Regulation 9).

Other legislation may require thorough examination and inspection, for example:

- **Pressure Systems Safety Regulations 2000**.
- **Control of Substances Hazardous to Health Regulations 2002 (COSHH)**.
- **Control of Lead at Work Regulations 2002**.
- **Control of Asbestos Regulations 2012**.

Dangerous Parts of Machinery

> **DEFINITION**
>
> **DANGEROUS PARTS OF MACHINERY**
>
> Those parts of a machine which may cause personal injury, including moving parts, sharp edges, etc.

Specific risks include dangerous parts of machinery. There is a specific duty on employers to take effective measures to prevent contact with these dangerous parts.

Such machinery should stop if persons enter the danger zone, i.e. where a person may be exposed to risks to their health and safety from contact with dangerous parts of the machinery.

Emergency Operations Controls

Machine and emergency control requirements are specified in **PUWER**.

Starting and Changing Operating Conditions

Where appropriate, all work equipment must be provided with one or more controls for:

- Starting the work equipment (including re-starting after a stoppage for any reason).
- Controlling any change in the speed, pressure or other operating conditions of the work equipment.

Stop Controls

Where appropriate, work equipment must be provided with one or more readily accessible stop controls, the operation of which will bring the work equipment to a safe condition in a safe manner. Activating the stop control should override all other controls.

Emergency Stop Controls

Emergency stop controls are emergency devices and should **not** be:

- used routinely to stop machines; or
- considered as alternatives to guarding the machinery.

BS 3641-2:1980 requires emergency stops to have a red mushroom-head, push-in button against a yellow background. Normally they should be of the lock-in type and re-setting the stop should not re-start the machine.

Emergency stops should be:

- Located at each workstation.
- Prominently displayed.
- Easy to reach.

Stability

Any item of work equipment, or any part of work equipment, that should be used in a fixed position, must be secured to prevent movement.

Emergency stop button

This may be by bolts, clamps, ties, etc. or, in the case of mobile equipment such as cranes, by the use of stabilisers, counterbalances and outriggers.

Lighting

Suitable and sufficient lighting, which takes account of the operations to be carried out, must be provided at any place where a person uses work equipment.

Additional lighting, over and above general lighting levels, may be needed:

- In respect of particular areas of the machinery, e.g. where access to dangerous parts is required.
- For particular operations, e.g. maintenance.
- Where dangerous operations may be carried out, e.g. excavation, demolition or working in the dark.

9.1 General Requirements for Work Equipment

Markings and Warnings

All controls for, and hazards of, work equipment must be clearly visible and identifiable by appropriate marking. Examples include:

- Operating instructions.
- Safe Working Load (SWL) marked on cranes, forklift trucks, lifting chains, and tackle and ropes.
- Maximum and minimum speeds for abrasive wheels, circular saws and bandsaws.
- Maximum and minimum size of components or workpieces.
- Hot, cold and abrasive surfaces.
- Ejection hazards, e.g. fumes, swarf, sparks, dust, etc.
- Hazard-specific, such as 'Radioactive'.
- General hazards, such as 'Mind Your Head'.

Markings can comprise words, letters, shapes or pictograms.

Specific visual or audible warning devices may be required to alert people to danger, for example:

- Flashing lights (beacons) on equipment.
- Fault lights on control panels.
- Audible reversing alarms on vehicles.

Visual warnings are limited - they rely on people looking in that particular direction. Audible warnings may not be suitable in a noisy working environment, especially if employees are wearing ear defenders, or for workers with a hearing impairment.

Clear and Unobstructed Workspace

The space within which workers are required to use tools and operate machinery should be enough to allow complete freedom of movement and to perform all the necessary operations in a safe manner. In particular:

- Space around machinery in use should allow clear separation from passing traffic and for the storage of tools and work-in-progress.
- The operator of any control should be able to see from the control position that no person is in a place where they would be exposed to any risk to their health or safety.
- Systems of work should be established to ensure that no person is in a place where they would be exposed to a risk to their health or safety as a result of the work equipment starting.

Where this is not reasonably practicable, physical barriers or floor markings may be needed to ensure an unobstructed workplace.

STUDY QUESTIONS

1. Define:
   (a) 'Work equipment'.
   (b) 'Use' (of work equipment).
2. What does a UKCA (UKNI) mark signify?
3. What environmental considerations should an employer have regard for to ensure the safe use of work equipment?
4. In what situations is a planned programme of preventive maintenance required, and on what should it be based?
5. In what situations are workers required to receive training in the use of a particular piece of work equipment?
6. Outline the three main characteristics of an interlocked guard.

(Suggested Answers are at the end.)

9.2 Hand-Held Tools

Hand-Held Tools

IN THIS SECTION...
- When selecting hand-held tools, requirements for safe use, conditions and fitness, and suitability for task are carefully considered.
- The use (and misuse) of hand-held tools can present a range of hazards. Appropriate control measures are important to ensure safety, including the condition of the equipment and suitability for purpose.
- There must be procedures in place to deal with defective equipment in every workplace.

Considerations for Selecting Hand-Held Tools

General considerations are:

- Task to be performed.
- Frequency and duration of use.
- Ergonomic design.
- Environmental conditions.
- Power supply.
- In-built safety features.

Requirements for Safe Use

Simple precautions for the use of hand-held tools generally relate to their condition and suitability for purpose. In particular:

- Users should have training and instruction in the correct use of tools - the skilled workers can teach the less skilled.
- Tools must be suitable for the job and the environment, e.g. in flammable or explosive atmospheres, non-sparking tools are required, i.e. non-ferrous (e.g. beryllium copper alloys), which require regular inspection to ensure their safe use.
- Correct maintenance, e.g. ensuring that blades are sharp (dull tools may slip and cause contact injuries).
- Correct storage and cleanliness, to avoid tripping hazards.
- Regular checks by the user and spot-checks by management.
- The withdrawal or repair of defective tools.
- Appropriate PPE, e.g. goggles and gloves, must be worn to protect eyes, hands and arms while using hand tools.

Virtually all persons will use hand tools on a construction site

Condition and Fitness for Use

Appropriate control measures are important to ensure safety, including the condition of the equipment and suitability for purpose.

The use (and misuse) of powered hand tools can present a range of hazards. The condition and suitability for use of pneumatic drills and chisels; electric drills; disc cutters and cut-off saws; sanders; cartridge and pneumatic nail guns; and chainsaws, is particularly important.

Hand-Held Tools | 9.2

Suitability for Purpose and Location of Use

Employers should ensure that tools are used for the purpose and in the environment they are designed for. This also means selecting equipment in accordance with the particular conditions in which it is to be used. This includes not using:

- Electric tools in damp or wet environments unless they are approved for that purpose.
- Power-actuated tools in an explosive or flammable atmosphere.
- Iron or steel tools around flammable substances or in flammable/explosive atmospheres.
- Petrol-driven tools in or near confined spaces.

Hazards of a Range of Hand-Held Tools

The same hazards presented by manually powered hand tools are increased by the presence of the power source (and especially the cables) and the speed and force of the tool itself. Injuries tend to be more severe and therefore the risk is higher.

General hazards include:

- Electrical problems - resulting in shock or burns from arcing or fire.
- Fuel spillages and fire from flammable fuel-powered tools.
- Noise and vibration from prolonged use.
- Manual handling problems may occur with larger tools.
- Puncture wounds from materials or tool parts.
- Entanglement in moving parts such as drive belts and pulleys.
- The emission of dust and other splinters or fragments.
- Tripping on cables and hoses powering the tools.

Some **equipment-specific hazards** include:

Pneumatic drills/chisels	High levels of noise and vibration. Manual handling issues carrying and holding them. Potential for impact injuries.
Electric drills	Risk of shock and burns. Entanglement in drill bit. Puncture by drill bit. Swarf flying off.
Disc cutters	Noise and vibration. Contact with high-speed disc causing entanglement, severe cuts and impact injuries. Ejection of debris and dust.
Cut-off saws	Noise and vibration. Contact with high-speed disc or blade causing severe cuts and impact injuries. Ejection of debris.
Disc, belt and orbital sanders	Noise and vibration. Creation of high levels of dust. Potential for contact with sanding belt or pad, causing entanglement, impact injuries and cuts.
Cartridge and pneumatic nail guns	Impact noise. Potential for puncture injuries from penetration of nails. Explosive risk of cartridges if mishandled or stored wrongly, and from misfires.

9.2 Hand-Held Tools

Non-Powered Tools

Hammers and Mallets
Avoid split, broken or loose shafts and worn or chipped heads. Make sure the heads are properly secured to the shafts.

Files
Files should have a proper handle. They should never be used as levers.

Chisels and Bolsters
The cutting edge should be sharpened to the correct angle. Do not allow the head of cold chisels to spread to a mushroom shape – grind off the sides regularly. Ensure the hand guard is in place.

Cutting Tools (Wire Cutters, Bolt Cutters, Knives, Pliers and Snips)
Use only cutting tools that are sharp and in good condition. Always cut away from the body and face. If a sharp tool is dropped, workers should be taught not to try to catch it. Make straight, even cuts without rocking, prying or twisting the tool.

Screwdrivers
Never use them as chisels and never use hammers on them. Make sure that they are the correct type and size.

Spanners
Avoid splayed jaws. Scrap any that show signs of slipping. Have enough spanners of the right size. Do not improvise by using pipes, etc. as extension handles.

Hand Saws
Avoid using hand saws that are too big or too small for the job. Inspect them thoroughly for any signs of damage. Do not use the hand saw if the handle is loose, the blade is bent or broken, or if there are missing blade teeth. Ensure the material is held firmly in place by using a vice or a clamp. Wear thick gloves if holding small items.

Control Measures for Safe Use
Because power tools have a greater risk of causing injury, employers need to ensure more stringent precautions are taken than for hand tools:

- Tools must be suitable for the job and the environment.
- Manufacturers' instruction booklets and company procedures should be followed.
- Users:
    - Must be trained in safe use and care of the tools.
    - Should be properly supervised in safe use.
    - Should check all tools before each use.
- Management should ensure formal checks are done.
- Formal inspection and maintenance should be done to a schedule.
- All defects should be recorded and tools taken out of use until repaired or replaced.
- Ensure regular breaks are taken to avoid vibration injury.
- Use all appropriate PPE for the tools and the work being done.

Hand-Held Tools 9.2

Procedures for Defective Equipment

Defective or damaged equipment should be taken out of service and quarantined with a label indicating the item should not be used until repaired.

Inspection and any subsequent tests and repairs should be carried out by a competent person. A record of inspection of such equipment should be made and kept for the life of the equipment.

In addition to regular inspections, operators should be instructed never to use damaged or defective equipment. They should visually check equipment before use and withdraw any defective items from service until repaired.

TOPIC FOCUS

Checks on **portable electrical equipment** should include:

- Visual inspection of the mains cable for damage.
- Confirmation of the correct cable for the tool.
- External inspection of the plug.
- Internal inspection of the plug wiring (where appropriate) and fuse - check for the correct rating.
- Cable correctly clamped in the plug and in the tool.
- On/off switch correctly operating and not damaged.
- Outer case undamaged.
- Earth bond test if a metal case.

For **air-fed equipment**, checks should include:

- Air hose connections are properly clamped.
- Hoses are undamaged and do not leak.

STUDY QUESTIONS

7. (a) What are the risks when using hand-held tools?

    (b) What are the additional risks when using portable power tools?

8. What procedures should be followed after a defect is reported in an item of work equipment?

(Suggested Answers are at the end.)

9.3 Machinery Hazards and Control Measures

Machinery Hazards and Control Measures

IN THIS SECTION...
- There are mechanical and non-mechanical hazards associated with machinery and many forms of harm that may result from them including crushing; shearing; cutting or severing; entanglement; drawing-in or trapping; impact; stabbing or puncture; high-pressure fluid injection; and friction or abrasion.

Consequences as a Result of Contact with Hazards Identified in ISO 12100:2010

Hazards arising from the operation of large, powered machinery may be divided into:

- **Mechanical hazards** - arising from the interaction of people with the machine itself.
- **Non-mechanical hazards** - associated with the use of machines, in respect of the environment, the materials used and other aspects of operation.

(These are described in BS EN ISO 12100:2010 *Safety of machinery*).

Mechanical Hazards

> **TOPIC FOCUS**
>
> There are six general factors about machines and the way in which people come into contact with them which cause specific **mechanical hazards**:
>
> - Shape of the machine - sharp edges, angular parts, etc., which may be a hazard even if not moving.
> - Relative motion of machine parts in relation to a person.
> - Mass and stability of the machine, parts of it, or the workpiece.
> - Acceleration of moving parts of a machine (or the workpiece), either under normal conditions or if something breaks.
> - Inadequate mechanical strength of a machine or part of it.
> - Potential energy of elastic components which may be translated into movement.

We can identify specific types of mechanical hazard arising from these general factors, each of which has the potential to cause serious personal injury or death:

- **Crushing**

    The body is trapped between two moving parts or one moving part and a fixed object, e.g. a vehicle reversing to a loading dock, or a tipper truck body dropping.

Crushing can occur beneath a tipper body

Machinery Hazards and Control Measures — 9.3

- **Shearing**

    A part of the body (commonly fingers) is trapped between two parts of a machine, one of which is quickly moving past the other. The effect is like a guillotine, shearing off the trapped body part.

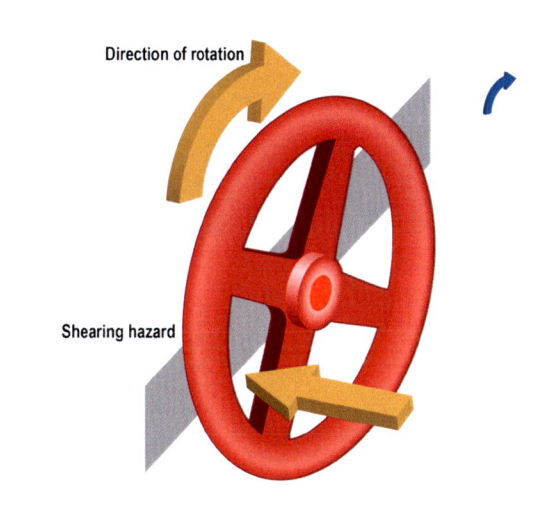

Shearing - a finger put through the spokes of this wheel will be sheared off

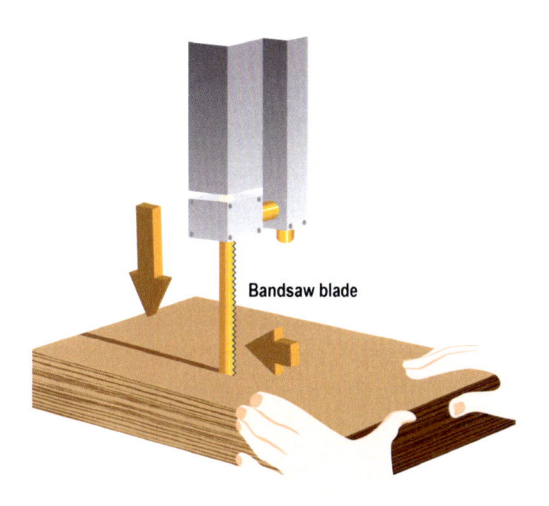

- **Cutting or Severing**

    A moving, sharp-edged part (a blade) is touched.

Cutting or severing - if the hands come into contact with the moving blade severe laceration or amputation will occur

- **Entanglement**

    Loose items of clothing or hair get caught on a rotating part of a machine and the person is drawn onto the machine.

Entanglement - a trailing scarf becomes entangled in the machinery

9.3 Machinery Hazards and Control Measures

- **Drawing-In or Trapping**

    A part of the body is caught between two moving parts and drawn into the machine, e.g. at 'in-running nips' where a chain and drive sprocket meet.

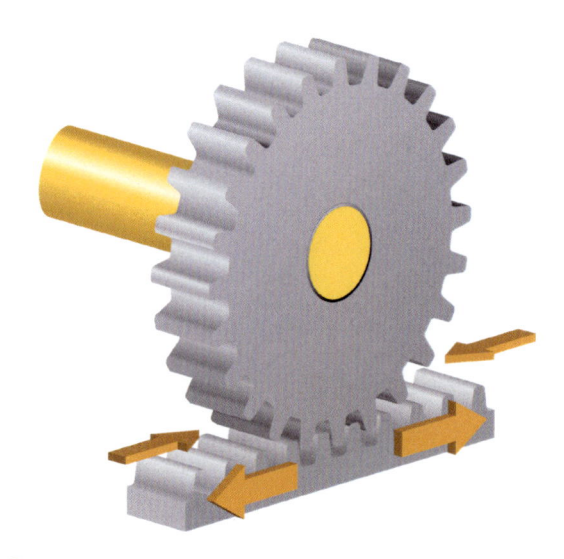

    Drawing-in or trapping - if the rollers are touched at the in-running nip point then the hand will be drawn in by the two rollers

- **Impact**

    A person is struck by a power-driven part of a machine when it moves, e.g. the swinging bucket of an excavator.

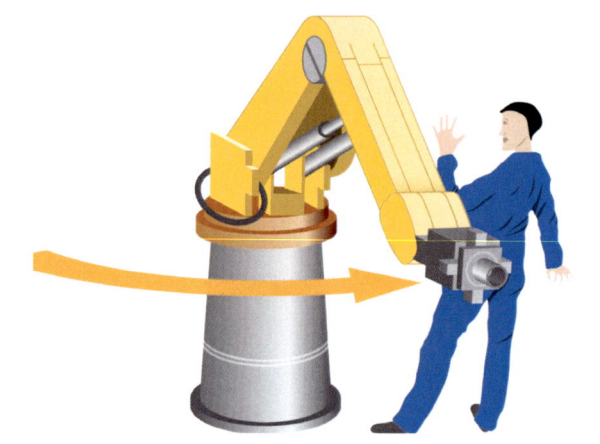

    Impact - the person is struck hard by the heavy and fast-moving industrial robot. A similar incident could occur as a result of walking in the path of an excavator bucket

- **Stabbing or Puncture**

    Sharp parts of a machine or materials in, or ejected from, a machine penetrate the body (e.g. metal swarf or nails from a nail gun).

Machinery Hazards and Control Measures | 9.3

- **Friction or Abrasion**

    Parts of the body (usually hands) making contact with moving equipment such as conveyor or drive belts.

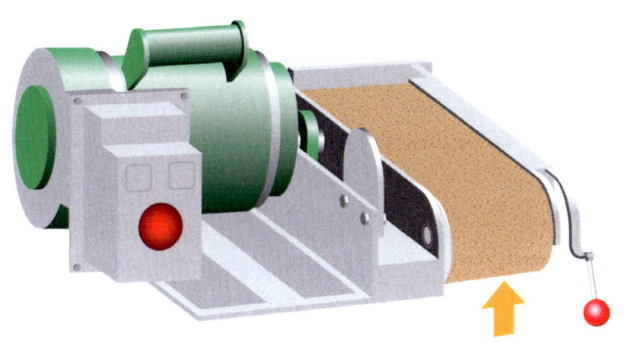

    Friction or abrasion - if the belt is touched whilst in motion then abrasion occurs

- **High-Pressure Fluid Injection**

    Hydraulic fluid ejected, often from a burst hose, can penetrate the skin. Hydraulic oil systems on telehandlers can be as high as 3200 psi. The tyre pressure of a family car is in the order of 50 psi. At such elevated pressures, the oil will penetrate the skin (often a small entry wound) but can cause significant damage or prove fatal.

Non-Mechanical Hazards

These are the hazards not directly arising from the moving parts of machinery. They arise from, or are created by, the power sources and the processes for which the machines are used.

TOPIC FOCUS

The non-mechanical hazards of machinery are:

- Electricity.
- Noise.
- Vibration.
- Hazardous substances.
- Ionising radiation.
- Non-ionising radiation.
- Extreme temperatures.
- Ergonomics.
- Slips, trips and falls.
- Fire and explosion.

9.3 Machinery Hazards and Control Measures

Hazards and Controls of a Range of Site Equipment

Workshop Machinery	Hazards	Control Measures

Strimmer	**Mechanical:**
	• High-speed cutting line.
	• Ejection of debris from machinery while cutting.
	• Handling of sharp items of litter and items with biohazard risks.
	Non-mechanical:
	• Slips and trips due to uneven or slippery ground.
	• Manual handling of equipment.
	• Contact with moving parts of machinery.
	• Contact with fuel and oils and plant materials.
	• Noise from machinery operations.
	• Exposure due to weather conditions.
		• Ensure that the item of strimmer is suitable for the task.
		• Ensure only trained and competent persons use strimmers.
		• Ensure the strimmer is in good order, including safety harness and blade.
		• Check all guards are present and correctly fitted.
		• Check the area to be treated for wire, glass or other hazards.
		• Consider the proximity of people and property, and flying debris.
		• Always use the correct PPE:
		  – Face shield.
		  – Ear defenders.
		  – Head protection.
		  – Safety footwear.
		  – Suitable overalls and gloves.
		• Check that the terrain is safe.
		• Follow manufacturers starting and operating procedures.
		• Take care when re-fuelling:
		  – allow engine to cool,
		  – avoid spills,
		  – wipe up spilt fuel and oil.
		• Switch off machine if approached by other persons.
		• Disconnect spark plug prior to carrying out any maintenance operation.

Chainsaw	**Mechanical:** • Entanglement. • Lacerations from rotating teeth. • Eye/face injury from flying debris. **Non-mechanical:** • Noise and vibration. • Sawdust. • Manual-handling injuries. • Slips, trips and falls. • Injury from falling branches.	• Specific training. • Two-handed operation with safety trigger. • Eye and hearing protection; anti-cut (padded) trousers, gauntlets, face visor. • Good housekeeping.

Machinery Hazards and Control Measures | 9.3

Workshop Machinery	Hazards	Control Measures
Cement mixer	**Mechanical:** • Entanglement with rotating drum or inner blade. • Drawing-in at nip between drive motor and drive mechanism. • Crushing between drum and drum stop when tipping. • Friction or abrasion on contact with moving drum. **Non-mechanical:** • Electricity. • Ergonomics from handling during loading. • Health hazard from inhalation and skin contact with cement/cement dust. • Noise. • Exhaust fumes (can cause problems if they enter trenches, excavations or any confined spaces).	• Set on level ground and secured against movement. • Fixed guards to motor and drive mechanism. • Routine inspection and portable appliance testing. • Reduced voltage to 110V and connection through Residual Current Device (RCD), a fast acting trip. • Use heavy duty cables or cable covers. • Use restricted to trained operators only. • Hand protection (impervious gloves) and dust/splash-resistant goggles; dust mask. • Place materials close to mixer to reduce handling. • Ensure mixer is not overloaded and manufacturer's instructions are followed. • Do not leave mixer unattended while mixing.
Bench-mounted circular saw	**Mechanical:** • Cutting on contact with the blade. • Entanglement with drive motor. • Drawing-in at nip between motor and drive belt. • Ejection of workpiece during cutting. **Non-mechanical:** • Electricity. • Noise and vibration-related health effects (e.g. vibration white finger). • Health hazard from inhaling wood dusts (breathing difficulties, asthma, etc.).	• Fixed guard fitted to motor and bottom of cutting blade. • Adjustable top guard fitted above blade. • Riving knife fitted behind blade (stops timber pinching shut on the saw blade after it has been cut - which can cause the timber to kick back towards the operator). • Hearing protection. • Vibration-insulated gloves or time-limitation. • Impact-resistant eye protection (goggles). • Extraction, exhaust ventilation, dust collector or face mask may be necessary. • Routine inspection and electrical testing. • Restrict use to trained operators only.

9.3 Machinery Hazards and Control Measures

Workshop Machinery	Hazards	Control Measures
Compressor	**Mechanical:** • Contact with moving parts (drive belts). • Entanglement and drawing-in - drive belts and pulleys. **Non-mechanical:** • Noise. • Vibration. • High-pressure fluid injection (compressed air). • Explosion of air receivers or pipework. • Health hazards from lubricants, air contaminants and dusts. • Fumes from exhaust gases. • Fire/explosion risks from refuelling. • Slips due to oil spills. • Burns from hot surfaces.	• Fixed guards around drive motor and belts. • Pressure relief valves and regulators. • Bursting discs in critical pressure locations. • Written scheme of examination for pressure systems. • Electrical testing and regular maintenance. • Hearing protection. • Use by skilled operator only.
Plate compactors and ground consolidation equipment (e.g. pedestrian-operated vibration rollers, vibrating compactor plates, wacker plates, etc.)	**Mechanical:** • Crushing in and beneath the equipment and on slides. • Impact/collision with the moving machines. **Non-mechanical:** • Manual handling injuries from handling the machines, e.g. foot injuries. • Health hazard from inhalation of dust/fumes. • Noise and vibration. • Fire and explosion from refuelling.	• Safety footwear and abrasion-resistant gloves. • Safe operating procedures. • Hearing protection. • Limit operator time to reduce fatigue, vibration and handling stresses. • Use restricted to trained operators only.
Road surfacing (e.g. rotary driers, batch heaters, mixers, bitumen and asphalt tanks) **and marking equipment** (lining equipment uses hot polymers and rolling machines to lay the painted lines on the road surface)	**Mechanical:** • Impact injuries from collision with moving machinery. • Entrapment beneath wheels/tracks. • Entrapment or drawing-in at material discharge rollers. • Entanglement. **Non-mechanical:** • Burns from hot polymers, bitumen and asphalt. • Health hazard from inhalation of fumes and smoke. • Fire and explosion risk from fuel/refuelling.	• Awareness of moving vehicles. • Side guards along material discharge rollers. • Overalls, heat-resistant gloves and respirators. • Safe storage and refuelling procedures. • Use by trained operatives only.

Machinery Hazards and Control Measures — 9.3

Workshop Machinery	Hazards	Control Measures
Electricity generators (petrol- or diesel-powered generators)	**Mechanical:** • Drawing-in to moving parts/entanglement. **Non-mechanical:** • Noise. • Petrol or diesel fumes. • Hazards associated with refuelling and storage of fuel. • Electricity.	• Covers and guards over moving parts. • Safe refuelling system and safe storage of fuel on site. • Safe system for connecting into electrical system to avoid electrocution. • Hearing protection when in close proximity. • Safe location to avoid exposing others to exhaust fumes, e.g. away from excavations and confined spaces.
Drones	**Mechanical:** • Collisions with people, buildings and other aircraft. • Collisions leading to property damage, personal injury, and loss of life in extreme cases. **Non-mechanical:** • Electricity. • Manual handling. • Hot surfaces and chemicals (if powered by a combustion engine)	• Keep the drone in clear sight • Do not fly higher than 120m. • Stay clear of aircraft and airfields • Do not fly closer than 50m to people. • Do not fly over a built-up area. • Obstacle-avoidance sensors. • Propeller guards. • Competent pilot.
Driverless vehicles	**Mechanical:** • Software anomalies resulting in crushing or impact. **Non-mechanical:** • Electricity (power source). • Chemical (battery acids.) • Non-ionising radiation (lasers).	• Avoid trapping locations. • Guards. • Wi-fi override. • Sensors to detect movement in front of the vehicle.

9.3 Machinery Hazards and Control Measures

Control Measures and Basic Requirements For Guards and Safety Devices

The subject of guards and safety devices was dealt with in detail earlier in this element.

There are six main requirements that all guarding systems should meet. They should:

- **Be Compatible with Process**

    During operation and maintenance, protection devices must cause the minimum interference with the function of the machine while still providing protection against the hazards.

- **Be of Adequate Strength**

    Protection devices must be:

    - Of sufficient strength to withstand impact from the hazard.
    - Appropriate to the extent of the risk.
    - Compatible with the working process.

- **Be Maintained**

    All guards and safety devices must be regularly checked and maintained in proper working condition. This can range from operator checks to regular inspections and servicing by trained staff. Records may need to be kept of the maintenance work.

    Any faults should be reported at once and repairs carried out as soon as possible.

- **Allow Maintenance Without Removal**

    It is possible to carry out certain maintenance procedures on machinery without the need for removal of any guards, e.g. maintaining oil or coolant levels, replacing filters, clearing air lines, and general machine cleaning (such as removal of cuttings, swarf).

- **Not Increase Risk or Restrict View**

    Many guards bring their own risks, e.g. certain types of screen may restrict the view of the machine. A balance must be struck between the benefits the guards offer and any problems they may bring about - the benefits nearly always outweigh the problems.

- **Not be Easily Bypassed**

    It should not be possible to operate the machine without the protection device in place. Special tools may be required to unlock perimeter fences around machines and gain access to a restricted maintenance area.

STUDY QUESTIONS

9. What are the general factors about machines and the way in which people may come into contact with them which cause the specific mechanical hazards in any situation?
10. What are drawing-in injuries?
11. List the non-mechanical hazards arising from the use of machinery.

(Suggested Answers are at the end.)

Working Near Water

9.4

Working Near Water

IN THIS SECTION...

- Precautions when working over or near water include warnings; scaffolding and temporary working platforms; buoyancy aids and safety boats (used as a last resort); platforms and gangways; good housekeeping; illumination; first-aid equipment; protective clothing and equipment; life buoys and rescue lines; safe operating procedures; and rescue procedures for use if other protection fails.
- Some specific work environments and equipment require additional controls, e.g. ladders, work in poor weather.

Additional Appropriate Control Measures

There are a number of precautions which can contribute to the safety of individuals and groups of people by preventing them from falling into water, or providing aid if they do fall in.

Warning Notices

These should be erected at all edges and boundaries near water and set so that they are easily seen by anyone approaching.

Scaffolds and Temporary Working Platforms

These should be erected by qualified, competent persons and inspected according to the regulations - prior to work, in inclement weather, and weekly. They provide the best method of ensuring safe working over water. The scaffold should be designed and inspected for the task so that it is stable and of sufficient size for the proposed work, with double-height toe boards, double guardrails, and brick guards or nets. Boards should be lashed to prevent damage from high winds.

Buoyancy Aids

Life jackets or buoyancy aids (designed to keep the wearer afloat) must be worn where there is a risk of drowning when working on or near water, and at all times while working on boats:

- A **life jacket** will provide sufficient buoyancy to turn and support even an unconscious person face upwards:
    - Inflation is by means of the mouth or carbon dioxide cartridge.
    - The life jacket supports the head with the mouth and nose well clear of the water.
    - Some people are reluctant to wear life jackets as they find them bulky and restrictive of movement.
- **Buoyancy aids** are worn to provide extra buoyancy to assist a conscious person in keeping afloat:
    - They will **not** turn an unconscious person over from a face-down position.
    - They tend to be less bulky than life jackets.

Life jacket

The selection of either type will depend on an assessment of:

- Water conditions - temperature, current, tides.
- The work being undertaken.
- The protective clothing being worn.
- The proximity of assistance.
- A person's competency as a swimmer.

9.4 Working Near Water

Safety Boat

The safe transport of any person conveyed by water to or from their place of work is a requirement of the **Construction (Design and Management) (CDM) Regulations 2015**.

Passenger-carrying craft must:

- Not be overcrowded or overloaded.
- Be marked with the number of persons they are intended to carry and, where appropriate, the limits of operation.
- Be suitably constructed and maintained.
- Be inspected if they carry more than 12 passengers at any one time, and a worthiness certificate obtained.
- Carry life-saving and fire-fighting appliances/equipment appropriate to their size.
- Be under the control of a competent person.

Communication is extremely important and should be in place during any work activity on or in the vicinity of water and in the event of an emergency situation arising.

Platforms and Gangways

Where platforms or gangways are erected, these must comply with the requirements of **CDM**. The decking boards should be safely secured with additional handholds. Working platforms, e.g. barges or pontoons, must be properly constructed, sufficiently stable to avoid tipping under loads, and have good anchorage and ballasting.

Ladders

Any ladders used for access must meet all usual requirements and:

- Be of sufficient length.
- Extend at least five rungs above a stepping-off point.
- Be securely lashed to prevent slipping.

Permanently fitted ladders over water are fitted with safety hoops.

Landing places must be provided every nine metres if longer access to the scaffold or platform is required.

Housekeeping

Good housekeeping on scaffolding, platforms and gangways is essential in preventing tripping hazards. All tools, equipment and rubbish should be stored away, stacked safely or disposed of. Any contaminants, e.g. bird droppings, slime, oil or grease, should be cleaned off or treated to prevent slips, injuries and minimise fire hazards.

Illumination

Illumination of the workplace is essential for night work and at all times in shafts, dark corners and stairways to avoid the possibility of shadows and glare. Illumination should always include the immediate water surface. Spotlights may be used to help locate a person in the water. Navigation lights and foghorns may also be required at working places in, on or near water and a check should be made with the appropriate authorities regarding requirements.

Weather Conditions

Rain, wind, fog, sea-mist and inclement weather are potential dangers during any work in the vicinity of water. The local weather forecast should be obtained and employees informed prior to each day's work or shift.

Working Near Water 9.4

First-Aid Equipment

First-aid facilities, appropriate first aiders and/or an appointed person should always be available (depending on the nature of the construction work).

The facilities/equipment available should:

- Be readily accessible and may be situated on pontoons, barges and near all possible landing places.
- Include portable equipment for resuscitation and transporting any casualties to the main working area over water and to normal landing places.

Protective Clothing and Equipment

- **Safety helmets** must be worn at all times, as anyone struck on the head and then falling into water is at a particular risk of drowning.
- **Footwear** with non-slip soles should be worn, while rubber and/or thigh boots should be avoided due to them filling with water, which could result in the wearer being dragged under water.
- **Safety harnesses** and **safety belts** are permitted under the **WAH Regulations** where it is not possible to provide a standard working platform or safety net, provided that they are always worn and always secured to a safe anchorage.

    Types include chest harness, full-body harness, safety rescue harness, etc. and these must be properly selected for a particular use.

    Operatives must be trained and instructed in their use.

Lifebuoys/Rescue Lines

- **Lifebuoys**
    - Approved lifebuoys, with rescue lines attached, should be set at conspicuous places along or near the water's edge.
    - Lifebuoys are normally 760mm in diameter with a 30m lifeline attached.
    - Lifebuoys are made of cork with canvas covering, or of polyurethane foam with a rigid PVC cover, both effective in salt or freshwater.
    - During any night work, self-ignition lights on the lifebuoys should be in use.
- **Rescue Line**
    - A standard rescue line incorporates 25m of line in a canvas bag with a small flotation chamber. The end of this line is held, while the bag is thrown towards the casualty.
    - Another type of rescue line involves a 40m line inside a capsule - sufficiently small to be carried or mounted in a cabinet. The line and the capsule will float, enabling the casualty to grab the line.

Potential users require regular training and instruction in the use of this equipment, and regular checks should ensure that lifebuoys and rescue lines are still in their proper place and that they are intact and not in need of repair, e.g. as a result of vandalism or other interference.

Safe Operating Procedure

To ensure work over or near water is carried out safely, it is vital that:

- A risk assessment and method statement is carried out for the task. This is usually reinforced with the issue of a permit to work.
- Continual checks are made to ensure that no one is missing.
- No lone working occurs.
- Operatives work in pairs so that there is always one to raise the alarm.
- Appropriate training is given to all personnel in emergency procedures.

9.4 Working Near Water

Rescue Procedure

Rescue procedures should be practised on a regular basis and should have:

- A procedure for raising the alarm (on site and off site).
- A drill to follow to get a rescued person ashore.
- Provision of first aid/resuscitation.
- A routine for getting persons to hospital for check-up following immersion in water (possibly polluted) or for treatment for an injury.

Various circumstances may combine to make a straightforward lifting operation out of the water impossible. Personnel should be trained and instructed in safe rescue procedures, especially if a casualty is injured, too heavy, fully clothed, or in a state of panic.

Regular rescue drills and procedures should take place.

STUDY QUESTIONS

12. What additional precautions may be taken when working over or in the vicinity of water?

13. What protective clothing and equipment is used when working over or in the vicinity of water?

(Suggested Answers are at the end.)

Summary

This element has dealt with the hazards associated with work equipment used in construction - on site, in the administrative environments and when working near water.

In particular, this element has:

- Explained the general requirements for work equipment including:
  - The suitability of work equipment and conformity with relevant standards and UK requirements.
  - Restrictions on work equipment which presents specific risks to health and safety.
  - The provision of information, instruction and training, particularly regarding dangerous parts of machinery.
  - The maintenance of work equipment, including planned preventive maintenance.
  - The requirements for examination and inspection.
  - The importance of operation and emergency controls.
  - Environmental and other considerations, such as the importance of stability; lighting; marks and warnings; and a clear and unobstructed workspace.
- Outlined the hazards and precautions for hand-held tools.
- Discussed the hazards of machinery including:
  - Mechanical hazards, such as crushing; shearing; cutting or severing; entanglement; drawing-in or trapping; impact; stabbing or puncture; high-pressure fluid injection; and friction or abrasion.
  - Non-mechanical hazards.
- Described the precautions and procedures required when working near or over water, including the prevention of drowning and appropriate control measures such as buoyancy aids and rescue boats.

Exam Skills

Question

Scenario
You are currently reviewing the toolbox talks that will be delivered to all operatives on site during the construction project. A few days ago, on a site safety tour you noticed a site operative operating a petrol disc cutter. Even though operated correctly you decide a toolbox talk should be developed covering the hazards and controls that personnel need to be reminded of whilst operating this piece of equipment.

Task: Work Equipment
Back in the office you decide to prepare a briefing document to be used at a toolbox talk on the subject of MSDs. In the briefing document you need to create two clear headings:

(a) Hazards associated with the use of petrol disc cutters. **(5 marks)**

(b) What control measures can be introduced to help reduce the risks whilst operating a petrol disc cutter. **(5 marks)**

(Total: 10 marks)

Approaching the Question

Now think about the steps you would take to answer this question:

Step 1   The first step is to **read the scenario carefully**. Note the question is focussing on a particular piece of work equipment and the associated hazards with it.

Step 2   Now look at the **task** - prepare some notes under the two headings: 'Hazards' and 'Control measures'.

Step 3   Next, consider the **marks** available. In this task, there are 5 marks available for the first part and 5 marks for the second part of the question. Tasks that are multi-part are often easier to answer because there are additional signposts in the question to keep you on track. In this task, you have to create a briefing document that is easy to understand, giving examples for each part can aid understanding. You will need to provide around 4 or 5 different pieces of information including examples for each part of this task, as some pieces of information may gain more than 1 mark as they will require additional detail. The headings will allow you to keep your response separate – this will also help the examiner when marking.

Step 4   **Read the scenario and task again** to make sure you understand the requirements and ensure you have a clear understanding of musculoskeletal disorders. (Re-read your study text if you need to.)

Step 5   The next stage is to **develop a plan** - there are various ways to do this. Remind yourself, first of all, that you need to be thinking about the causes of MSDs and the measures that can reduce the effects of them.

Exam Skills

Suggested Answer Outline

Hazards associated with the use of petrol driven disc cutters:

- Ejection hazards.
- Noise and vibration.
- Dust.
- Refuelling – petrol and fire.
- Cuts and abrasions.

Control measures to reduce the risks of operating a petrol driven disc cutter:

- Create a safe work area – protection of third parties.
- Wearing of PPE (gloves, safety boots, ear protection, mask and eye protection).
- SSW – refuelling of the disc cutter.
- Health surveillance.

Now have a go at the question yourself.

Example of How the Question Could be Answered

(a) *The hazards associated with the use of a petrol driven disc cutter would be ejection hazards created as the cutting-edge cuts through materials. Noise and vibration would be a significant risk caused by the motor and vibration from its use. Dust would be created from the material which could be inhaled. Refuelling of the disc cutter with petrol which could be ignited from the hot surface of the motor. Finally, cuts and abrasions could be sustained by coming into contact with sharp edges of materials and possible contact with the saw blade.*

(b) *Control measures that can be introduced to reduce the risk of injuries and possible ill health from petrol driven disc cutters could first ensure the work area is safe with sufficient space to operate the equipment and, in particular, away from other personnel who could be affected by the dust and ejection hazards. This could be done by placing barriers around the work area. The provision of PPE such as gloves to protect operative's hands and to keep them warm in cold conditions. This will help in reducing the effects of Hand Arm Vibration Syndrome (HAVS). Provision of ear protection will help reduce the risk of noise-induced hearing loss. The use of eye protection will provide protection from dust and ejection hazards and finally dust masks to prevent inhalation of dust particulates. The creation of a Safe System of Work (SSW) for refuelling of the disc cutter would ensure it is not refuelled when the equipment is still hot allowing it to cool down first.*

Reasons for Poor Marks Achieved by Exam Candidates

- Not following a structured approach for the briefing document; failing to provide information on the two subject areas.
- Not expanding the answer beyond a few words as opposed to giving a sentence of explanation.

Element 10

Electricity

Learning Objectives

Once you've studied this element, you should be able to:

1. Outline the hazards and risks associated with the use of electricity in the workplace.

2. Outline the control measures that should be taken when working with electrical systems or using electrical equipment.

3. Outline the control measures to be taken when working near or underneath overhead power lines.

Contents

Hazards and Risks	**10-3**
Risks of Electricity	10-3
Control Measures	**10-8**
Protection of Conductors	10-8
Strength and Capability of Equipment	10-8
Protective Systems - Advantages and Limitations	10-8
Use of Competent People	10-10
Use of Safe Systems of Work	10-11
Emergency Procedures	10-13
Inspection and Maintenance Strategies	10-13
Control Measures for Working Underneath or Near Overhead Power Lines	**10-18**
Legal Requirements for Working Near Power Lines	10-18
Preventing Line Contact Accidents through Management, Planning and Consultation	10-18
Use of Barriers to Establish a Safety Zone When Working Near Overhead Lines	10-19
Means of Safely Passing Underneath Overhead Lines	10-19
Key Emergency Procedures for Contact with an Overhead Line	10-20
Control Measures for Working Near Underground Power Cables	**10-22**
Planning the Work	10-22
Using Cable Plans	10-22
Use of Service Locating Devices	10-23
Safe Digging Practices	10-23
Use of Appropriate Tools, Locating Devices and Route Planning When Undertaking Excavation Work	10-24
Summary	**10-25**
Exam Skills	**10-26**

Hazards and Risks

IN THIS SECTION...

- The hazards and risks that can arise from electricity include:
  - Electric shock and its effects on the body, including factors influencing severity of a shock: voltage, frequency, duration of exposure, the body's resistance and current path.
  - Burns from contact with electricity, both direct and indirect.
  - Electrical fires and their common causes.
  - The conditions and practices likely to lead to accidents when using workplace electrical equipment including portable items, such as unsuitable equipment, inadequate maintenance and using defective apparatus.
  - The secondary effects of electric shock, including falls from height.
  - The effects of working near overhead power lines, underground cables and on mains electricity supplies.

Risks of Electricity

Electric Shock and its Effects on the Body

This occurs when a person touches a live conductor and current passes through their body, using the body as a conductor. The current travels from the point of touching (a finger, hand, etc.) to a point in contact with the ground (an earth), often a foot or knee.

Electric shock can have different effects on a person. The amount of current (in amps) passing through the body will determine what these effects will be.

Current (mA) Flowing Through the Body	Effect of the Shock
0.5 - 2	Threshold of sensation.
2 - 10	Tingling sensations, muscle tremor, painful sensations.
10 - 60	Muscle contractions, inability to let go, inability to breathe.
60 and above	Ventricular fibrillation (see below), cardiac arrest, extreme muscle contractions, burns at contact points and deep tissues.

In the above table, the current is AC and is measured in milliamps (one thousandth of one amp) (1mA = 0.001A).

To explain these effects further:

- At very low current flow (less than 0.5 - 2mA) no sensation is felt.
- Between 2 and 10mA current starts to flow through the body and stimulates muscle contraction, felt as trembling and some pain. You maintain muscle control and can let go.
- Between 10 and 60mA more severe muscle contraction occurs, so strong that you may not be able to let go of the live conductor. Muscles of the abdomen and rib cage may also contract making breathing difficult (which means you can't call for help). Asphyxiation may occur. Alternatively, the shock may cause a massive muscle contraction of big muscle groups that throws the person off their feet (hopefully away from the live conductor).
- At current flows above 60mA Ventricular Fibrillation (VF) may occur, which puts the heart out of normal synchronised beating rhythm and makes it beat spasmodically. This usually leads to cardiac arrest.
- As current increases to 80mA VF is far more likely. Muscle contractions can be so severe as to break bones, and burns will appear at the points where the current enters and leaves the body (and also in the internal tissues where it has passed through). Death becomes more likely as current increases.

10.1 Hazards and Risks

What Affects Severity

- **Voltage** - Ohm's Law has shown us a simple relationship between voltage and current: the higher the voltage, the greater the current.
- **Frequency** - of the alternating current: the number of times it changes direction.
- **Duration** - the time in contact with the conductor is critical. For example, a current flow of 60mA for 30 milliseconds (30 thousandths of a second) is unlikely to cause severe injury. The same 60mA for just two seconds can induce VF and be fatal.
- **Current path** - the route the current takes as it passes through the body is critical. If it goes through the chest, it may directly affect the heart.
- **Resistance** - remembering the relationship in Ohm's Law between current and resistance, the higher the resistance, the lower the current. The skin provides most of the body's resistance, so **dry skin** will have a resistance of around 100,000 ohms. **Wet or damaged skin** reduces resistance to around 1,000 ohms - a very big reduction. Clothing between the skin and live conductor will also affect resistance.
- **Contact surface area** - the more skin (the larger the body area) in contact with a live conductor, the lower the resistance and the greater the injury.
- **Environment** - wet surfaces, humid air and metal surfaces will all influence the severity of a shock by reducing resistance.

Electrical Burns

People receive both external and internal electrical burns in two ways from electrical accidents:

- **Direct burns** - where electrical current causes overheating as it passes through the skin and the internal tissues of the body. External burns may occur at the sites of entry into and exit from the body, and these will be full skin thickness. Internal tissue burns can be very severe and may prove fatal.
- **Indirect burns** - do not occur from electrical current passing through the body, but from an electrical current causing something to overheat and explode, e.g. dropping a metal spanner onto a high-voltage cable. This causes a short circuit, which results in a flash of radiant heat and explosion of molten metal.

Common Causes of Electrical Fires

Fires need three elements in order to start - a source of **heat**, combustible material as **fuel**, and **oxygen**. Electricity may provide the source of heat by:

- **Sparking** - the generation of electrical sparks or arcs between an uninsulated or poorly insulated conductor and another, earthed conductor. Where such sparking occurs in the presence of flammable gases or in flammable or explosive atmospheres, then fire or explosion may result.
- **Overheating of conductors** - this may be due to:
    - Poor or inadequate insulation.
    - Excessive resistance within the conductor.

The sparking or overheating may come about as a result of:

- Faulty electrical equipment.
- Electrical system or equipment being overloaded.
- Protection systems incorrectly rated (e.g. fuses).
- Equipment abused or misused.
- Sparks or heat from normal operation, e.g. fans and drive motors running hot.
- Charging electrical equipment.

Hazards and Risks

Arcing (sometimes called 'flashover') is where electricity jumps across an air gap. In other words, it is the electrical bridging, through air, of one conductor with a very high potential, to another nearby conductor connected to earth. This occurs in a very limited way in some equipment, such as an electric drill. The higher the voltage of the equipment or circuit, the further an arc can jump. The main risks from arcing are:

- Electric shock from being struck by the arc.
- Direct burns from being struck by the arc.
- Indirect burns from the:
  - Radiant heat given off by the arc.
  - Melting of any equipment struck by the arc.
- Damage to the eyes from ultraviolet (UV) light emitted from the arc.

Many fires and explosions are due, not to any direct fault in an item of electrical equipment, but to the wrong type of equipment being used in the wrong place, such as:

- An item not intrinsically safe used in an atmosphere known to be flammable or explosive.
- Equipment used in an atmosphere not known to be dangerous, e.g. after a spillage.

Short Circuits

When insulation becomes faulty, or if one conductor touches another, an unintended flow of current between two conductors occurs, providing an alternative path to a terminal with a larger potential difference than the neutral terminal, usually the earth. This is known as a 'short circuit' and is often the cause of fire.

Static Electricity

Another cause of fire and explosion is **static electricity**, which is unlike the battery and mains supplies we have already discussed. A build-up of electrons on poor electrical conductors or insulating materials will create static electricity. This can be experienced in the home, for example, when a person walks across a new carpet. The friction between the shoes and the carpet builds the potential, which is released through the hand touching, for instance, a door handle. This 'shock' in the fingers is the release of the static in a very short duration spark.

Static electricity results from a build-up of potential difference (voltage) between surfaces or materials as a result of friction between them. Such materials can be in almost any form - liquid, gas or solid - and include many raw and process materials, such as powders and granules. The static can be caused by the movement or rapid separation of the materials by friction (e.g. passing through delivery pipes) or induction while passing from one material (e.g. a road tanker) to another (e.g. a storage tank or drum).

Static electricity can build up on people and materials, and can be discharged on contact. Sparks from equipment and people discharged in flammable atmospheres can be the cause of fire. One of the most common static discharges is lightning.

Workplace Electrical Equipment

Portable Electrical Equipment

Portable electrical equipment is any item or appliance connected by a flex and plug (or connected through a fused spur), so that it can be moved and used in different places.

There are risks associated with the use of workplace electrical equipment, including portable items:

- Portable electrical equipment accounts for almost a quarter of all electric shock injuries.
- In many cases, the equipment used is the wrong equipment, not suitable for the work.
- Often it is never maintained.
- Some items may still be used even though they are known to be defective.

Consider a mains-powered electric drill used in building refurbishment. It is often handled and transported without any secure means (e.g. a case or box) and:

- Dropped in the back of a vehicle.
- Lifted up and down ladders by the flex.

10.1 Hazards and Risks

- Dropped onto a wet floor or into dust and dirt.
- Rarely checked or inspected until it fails.
- Used by many different employees.

Poorly Maintained Electrical Equipment

Without routine inspection, maintenance and testing the integrity of electrical items cannot be assured. Simple defects are not spotted, leaving operators at risk of shock or other injury.

Use of Electrical Equipment in Wet or Flammable Atmospheres

All operators and other persons are at risk of electric shock, fire and explosion if items used in these areas are not intrinsically safe. This includes all fixed items (switches, lights, etc.) as well as portable items.

- Use in wet conditions can allow water or other liquids in the atmosphere or on the person's hands to enter the equipment and short-circuit it, causing electric shock to a person in contact with the equipment.
- Use in flammable or explosive atmospheres can allow stray current (in the form of sparks) from the electrical equipment to contact a flammable or explosive air mixture and ignite it.

> **TOPIC FOCUS**
>
> Causes of workplace electrical equipment accidents:
>
> - Using unsuitable equipment (e.g. using non-intrinsically-safe items in flammable atmospheres).
> - Using equipment in wet, damp or humid conditions.
> - Misuse, e.g. sticking wires directly into a socket rather than fit a plug.
> - Physical abuse, e.g. pulling the plug from a socket by its flex; carrying it by its flex; allowing the flex to be pinched, trapped or crushed.
> - Carrying out unauthorised repairs, e.g. taping up a split flex.
> - Continuing to use the equipment knowing it is faulty or defective.
> - Chemical damage to the flex or tool from harsh chemicals, such as corrosive wet cement.
> - Lack of routine inspection, maintenance or testing.

Secondary Effects

As well as any injuries caused directly by an electric shock, secondary hazards can arise, based on what the person was doing when they received the shock, or when a short circuit occurred.

Examples:

- A machine develops a fault and part of the machine or process becomes dangerous, putting the person at risk.
- A person is working on a ladder and the shock throws them from the ladder to the ground, or into the path of a moving vehicle.
- A person is working at height and falls or causes items of equipment/materials to fall to the ground, injuring someone else.

Work Near Power Lines

When working near power lines:

- Avoid where possible.
- Ensure pre-planning and consultation with service providers.
- Supplies to be isolated or diverted.

Hazards and Risks 10.1

- Use ground-level barriers, including goal-posts.
- Safety signs and correct clearance distances.
- Restrict access to certain types of equipment.
- Vehicles with long reach are controlled.
- All points that pass beneath cables should be signed and lit if work is to be carried on at night.

Contact with Underground Cables During Construction Work

Contact with underground cables usually occurs during excavation work and can expose operators to mains power cables and the potential for electric shock. The cables may be unmarked or unidentified on plans and site maps. When struck by an excavator (or even by a shovel when hand-digging), the outer cover of the cable can be penetrated allowing the inner live conductors to pass live electricity into the machine or shovel, electrocuting the operator.

Work on Mains Electricity Supplies

When working on mains electricity supplies, be aware of the following:

- Contact with live parts can cause shock and burns.
- Faults which could cause fires.
- Fire or explosion where electricity could be the source of ignition in a potentially flammable or explosive atmosphere, e.g. in a spray paint booth.

STUDY QUESTIONS

1. What is the main effect of electric shock on the body?
2. What is arcing and what risks does it pose?
3. Identify three risks associated with electricity.

(Suggested Answers are at the end.)

10.2 Control Measures

Control Measures

IN THIS SECTION...

- Precautions must be in place to prevent electrical incidents and to minimise effects should an incident occur. Such precautions include:
    - Protection of conductors.
    - Suitability and capability of equipment.
    - Protective systems, including fuses, earthing, isolation, double insulation, reduced or low voltage systems and residual current devices.
    - Use of competent people.
    - Use of safe systems of work including no live working and safe isolation, locating buried services and protection against overhead cables.
    - Having emergency procedures in place for an electrical incident, including first-aid treatment for shock and burns.
    - Inspection and maintenance of electrical systems, including user checks, formal and frequent inspection and testing, keeping records of inspections and tests, and portable appliance testing.

Protection of Conductors

Protection of conductors by covering them with insulating material is, in most cases, the primary safeguard to prevent electric shock. It will also prevent danger from fire and explosion. The electrical conductors in a system also need protection by being installed to an appropriate standard and safely earthed.

Strength and Capability of Equipment

Regulation 5 of the **Electricity at Work Regulations 1989** states:

"No electrical equipment shall be put into use where its strength and capability may be exceeded in such a way as may give rise to danger."

Protective Systems - Advantages and Limitations

Protective devices incorporated into electrical circuits or the equipment itself act mainly to cut off the electricity supply in the event of a fault and/or to reduce the current delivered to a person in the form of an electric shock.

> **TOPIC FOCUS**
>
> Various **protective systems** can be used for electrical equipment, such as:
>
> - Fuses - a weak link in the circuit.
> - Circuit breaker - a mechanical switch which automatically opens when the circuit is overloaded.
> - Earthing - a low resistance path to earth for fault current.
> - Isolation - cutting the power.
> - Reduced low voltage - so that less current flows during an electric shock accident.
> - Residual Current Devices (RCDs) - sensitive and fast-acting trips.
> - Double insulation - separating people from the conductors using two layers of insulation.
>
> Each of these protective systems has advantages and disadvantages; portable electrical appliances should undergo user checks, formal visual inspections, and combined inspection and testing to ensure electrical safety.

Control Measures 10.2

Protective System	Features	Advantages	Disadvantages
Fuses	Will prevent current overload. If the current passing through a fuse is too high (and exceeds the safe limit), the fuse melts and breaks the circuit.	Inexpensive and reliable; offer good protection for the item of equipment.	May not prevent electric shock. Easy to bypass.
Earthing	Protects users of electrical equipment by allowing fault current to safely flow to earth along a wire attached to the casing or chassis, rather than through a person who may touch it. Touching the casing will only result in minor shock.	Protects a person from fatal electric shock. Provides secondary protection to equipment (a large fault current flowing to earth will overrate the fuse or circuit breaker).	A poor or broken earth connection will prevent it from working properly, and can easily go undetected. Easy to disconnect and disable.
Isolation	Separation from and removal of the energy source, i.e. electrical power. Includes additional steps to prevent it being re-energised. Employed for safety purposes when work is to be carried out on electrical systems.	Very effective in ensuring people working on systems cannot be electrocuted by them.	By definition it is 'dead' so fault finding, testing, etc. cannot be carried out with the system isolated.
Double insulation ▢ This symbol is displayed on double-insulated equipment	Two layers of insulation between the live conductors and the equipment user. Eliminates the need for earth protection, so double-insulated items will only have a two-core flex - live and neutral.	Relies on insulation rather than the electrical system itself for safety.	Must be routinely visually inspected because there is no earth protection.
Reduced/ low voltage systems	The lower the voltage a system operates at, the lower the risk of injury (as voltage reduces, shock current reduces and severity of injury reduces). Step-down transformers reduce voltage in the UK from 240V to 110V or less for portable tools. Safety Extra Low Voltage (SELV) systems operating at 50V or less are available. Very low voltage systems operate as low as 12V.	Low voltages are inherently safer.	Inefficient at transmitting power and therefore cannot be used for many industrial applications.

10.2 Control Measures

Protective System	Features	Advantages	Disadvantages
Residual Current Devices (RCDs)	Specially designed to protect human life in the event of electric shock; sensitive to small current imbalances - break a circuit very quickly. An RCD constantly compares the amount of current flowing in the live and neutral lines. If an imbalance is detected it trips the circuit. RCDs (and Earth Leakage Circuit Breakers - ELCBs) can be: • Incorporated into electrical equipment (as part of the plug). • Stand-alone devices placed between a portable appliance plug and power socket. • Hard-wired into distribution systems such as the 'consumer unit' in a domestic house. RCDs should always be used on all portable appliances used outdoors.	Provide excellent protection for people in the event of electric shock. Quick and easy to reset.	Do not provide over-current protection (they are not a fuse). Have to be tested periodically, which often is not done. Can cause repeated circuit tripping if there is a fault, encouraging them to be disabled.

Use of Competent People

A number of areas in health and safety legislation require that work (including work on electrical systems and equipment), such as inspection and testing, is carried out by 'competent persons'.

> **DEFINITION**
>
> **COMPETENT PERSON**
>
> *"A person is accepted as competent where he has sufficient training and experience or knowledge and other qualities to enable him to properly do the task in question."*
>
> MHSWR 1999
>
> A competent person should be able to produce evidence of such competence.

For the purposes of electrical safety in examination and testing, 'electrically competent persons' must be qualified and trained to be aware of all aspects of safety relating to the examination and testing they are carrying out.

Where employees are not instructed on and trained in the safe implementation of safe systems of work for electrical systems and circuits, they should only work under the supervision of a competent person.

Use of Safe Systems of Work

Live Working

The **Electricity at Work Regulations 1989** (Regulation 4) requires that work on or near to an electrical system *"shall be carried out in such a manner as not to give rise, so far as is reasonably practicable, to danger"*. Again, such work includes examination and testing.

There is a strict restriction (Regulation 14) on working on or near live conductors unless there is no other option available. No live work shall carry on unless:

- it is unreasonable in all the circumstances for the equipment to be dead; and
- it is reasonable for the work to take place on or near the live conductors; and
- suitable precautions (including, where necessary, the provision of suitable protective equipment) have been taken to prevent injury.

All three conditions must be met in order for work on or near live conductors to be carried out. Otherwise, dead working should be the normal method of carrying out work on electrical equipment and circuits.

Safe Isolation

> **DEFINITION**
>
> **ISOLATION**
>
> *"...the disconnection and separation of the electrical equipment from every source of electrical energy in such a way that this disconnection and separation is secure."*
>
> **Electricity at Work Regulations 1989**

Adequate precautions must be taken to ensure that any electrical equipment made dead (so that danger can be prevented while work is carried out on it), cannot become electrically charged during that work. In other words, it cannot be switched on again during work. This is known as 'isolation'.

Safe isolation will require a competent person to:

1. Locate and identify the circuit or equipment to be isolated.
2. Verify the circuit or equipment is functional with an approved test device.
3. Identify a suitable means of isolation - mains switch, fuse board or distribution board.
4. Switch off the supply at the identified means of isolation.
5. Place locks in the means of isolation. A 'tag' should be fitted where more than one electrician may be working on the same system (each can use their individual lock).
6. Fit a warning sign to the means of isolation.
7. Use a test meter to 'prove' the circuit is dead.
8. Test the meter itself to show correct functioning.

The circuit or equipment should now be safe to work on.

On a construction site it is essential to also isolate and prove dead any secondary source such as generators.

10.2 | Control Measures

Locating Buried Services

Before digging or excavating, site plans should be checked to see if any buried services are in close proximity. Electronic locator devices can also be used to identify live electrical cables. When coming into proximity of a known buried cable, mechanical diggers should stop and hand-digging begin, proceeding with care until the cable is discovered. All cables, when revealed, should be marked.

Protection Against Overhead Cables

Avoidance Where Possible

Many fatalities and major injuries occur from contact with live overhead power lines. A sensible precaution at all times is to assume all overhead lines are live.

It is not necessary for there to be direct contact with an overhead line - a close approach to them can result in a flashover (arcing) taking place. The higher the line voltage, the greater the risk of this happening.

Work near to or beneath overhead power lines should always be avoided where possible.

Avoid working near overhead powerlines where possible

Pre-Planning and Consultation with Service Provider

Where work near or beneath the lines cannot be avoided, contact should be made with the owner (e.g. electricity supplier, railway operator, developer) to discuss the work involved. The purpose of this is to:

- Divert all overhead lines near the work site.
- Make the lines dead (isolate the supply) while work is in progress.
- Schedule the work if they can only be made dead for short periods of time.
- Adopt a safe system of work/procedure/precautions/permit-to-work system.

Three situations generally arise in construction at overhead power lines:

- Work in the vicinity of overhead power lines.
- Plant and equipment will pass under the lines.
- Work will be carried out beneath the lines.

The precautions required in each case are set out below. In all cases, work should be under the supervision of an appointed responsible person.

Ground-Level Barriers Working in the Vicinity of Overhead Power Lines

When working near overhead power lines barriers and bunting or flags may be used to signal safe zones (see later for more detail).

Plant/Equipment Passing Under the Lines and Restriction of Equipment and Vehicle Reach

When passing under the lines ensure that barriers, bunting and goal-posts are used to control the distance between plant and equipment (see later for more detail).

Emergency Procedures

First-Aid Treatment for Electric Shock

The first action should be to break any continuing contact between the victim and the current; in other words, switch off the power supply, because:

- The casualty may still be receiving a shock.
- High-voltage conductors may arc and strike the first aider.

If the power cannot be switched off, carefully push or pull the casualty away from the live part using non-conducting material such as wood or dry clothing.

The next steps are:

- Do not touch them unless the power source has been isolated or contact broken.
- Shout for help.
- Call for an ambulance.
- If power still on - push victim away from live contact with non-conducting material.
- Check breathing:
    - If breathing, place in recovery position.
    - If not breathing, apply Cardio-Pulmonary Resuscitation (CPR).
- Treat any obvious burns.
- Treat for physiological shock.
- Get medical attention (heart problems and internal burns may not be apparent to the casualty or a first aider).
- Stay with the person until help arrives.

First-Aid Treatment for Electrical Burns

When treating first-aid for electrical burns:

- Place a sterile dressing or pad of clean material over the burn, secure this with a bandage or tape. Avoid using blankets or towels as fibres may contaminate the wound.
- Do not burst blisters or remove any loose skin.
- Do not apply water, or any lotions or ointment.

Inspection and Maintenance Strategies

All electrical equipment and systems should undergo regular inspection and maintenance to ensure their safety. In many cases testing will also be necessary.

Where visual inspection shows that equipment is unsafe, it is to be taken out of service and repaired, or discarded and replaced.

10.2 Control Measures

> **TOPIC FOCUS**
>
> There should be an **inspection and maintenance strategy** in place. The basic requirements are:
>
> - Identification of the equipment which has to be maintained and where/how it is to be used.
> - Discouragement of 'unauthorised' equipment in the workplace.
> - Carrying out simple user checks for signs of damage, e.g. casing, plug pins and cable sheath.
> - Formal visual inspections carried out routinely by a competent person.
> - Periodic testing of equipment by a competent person.
> - Systems for the reporting and replacement of defective equipment.
> - Recording of all maintenance and test results along with the inventory of equipment in use.

There are several types of inspection and testing that might be appropriate for portable electrical appliances.

User Checks

All users of portable electrical appliances should themselves carry out a visual inspection of the item before they use it. This is especially important in harsher environments, e.g. an electric drill used on a construction site.

The user check does not involve any dismantling, but a careful visual inspection of the equipment to ensure the:

- Body of the plug is intact and secure.
- Outer sheath of the flex covers the inner cores from the body of the plug to the body of the appliance.
- Plug cable clamp appears tight.
- Flex appears fully insulated - no cracks, splits or severe kinks/pinches.
- Body of the appliance is intact.
- Appliance cable clamp appears tight.
- Plug and body of appliance have no obvious scorch marks.
- Plug and appliance are not:
  - Excessively soiled.
  - Wet.

Control Measures | 10.2

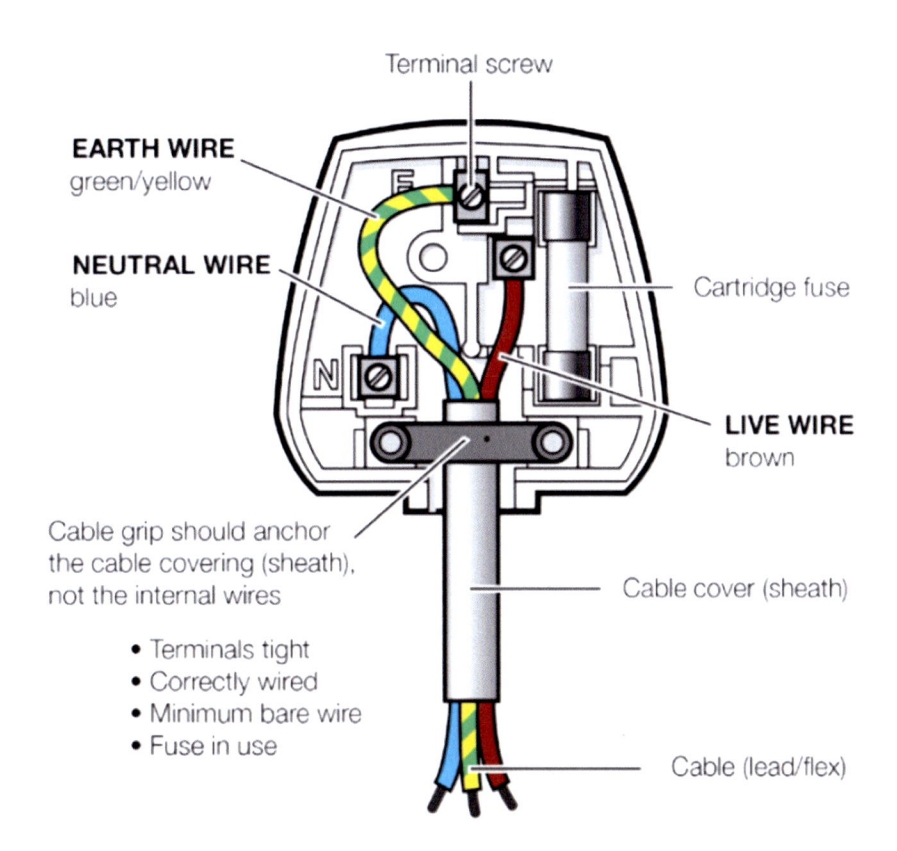

EARTH WIRE green/yellow

NEUTRAL WIRE blue

Terminal screw

Cartridge fuse

LIVE WIRE brown

Cable grip should anchor the cable covering (sheath), not the internal wires

- Terminals tight
- Correctly wired
- Minimum bare wire
- Fuse in use

Cable cover (sheath)

Cable (lead/flex)

A correctly wired plug: body of plug should be intact and secure. Source: HSG107 Maintaining portable and transportable electrical equipment (3nd ed.), HSE, 2013

Formal Inspection and Tests

There are two types of formal inspection - routine visual inspections and combined inspection and tests (portable appliance tests).

Inspection	Features	Advantages	Disadvantages
Routine visual inspections	Carried out by a competent person. Done at regular intervals. Go further than user checks and will include some dismantling of plugs and equipment.	Inexpensive. Trained person can do them easily and quickly.	Can miss things, e.g. deterioration of insulation and earth (leakage) faults.
Combined inspection and tests (Portable Appliance Testing (PAT))	Uses the visual inspection already mentioned, together with a Portable Appliance Testing (PAT) device, which carries out (and, in some cases, records) more detailed tests automatically.	Will detect unseen problems, e.g. deterioration of insulation and earth faults. Allows early removal of faulty items. Demonstrates legal compliance. Trends or patterns in faults may be spotted.	Requires a little more detailed knowledge of the equipment and testing equipment. Proves safety only at the time of testing. Does not ensure safe use or prevent misuse. Items may be missed and remain unsafe. May not be suitable for all equipment.

10.2 Control Measures

Frequency of Inspection and Testing

Various factors will influence how often checks are carried out. As an example, on a UK construction site, because of the conditions of the site and use of the equipment, a 110V hand-held power tool should be:

- Visually checked by its user once each week.
- Formally visually inspected once each month.
- Inspected and tested (PAT) once every three months.

Records of the formal visual inspections and tests should be kept to provide a history of condition and defects. It is good practice to attach a sticker to tested items showing the date of the next test.

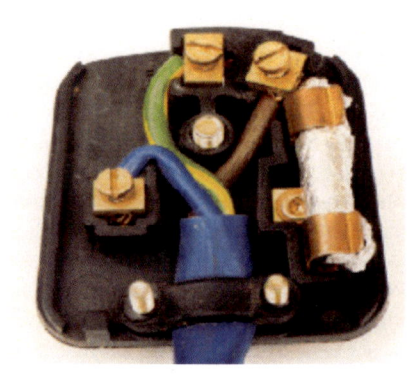

Formal inspection should uncover unsafe conditions, such as this fuse which has been disabled by wrapping it in tin-foil (Source: HSG107 Maintaining portable and transportable electrical equipment (2nd ed.), HSE, 2004)

TOPIC FOCUS

Factors that influence the **frequency of inspection and testing** include:

- Legal standards and codes of practice.
- Type of equipment and whether or not it is hand-held.
- Manufacturer's recommendations.
- Initial integrity and soundness of the equipment.
- Age of the equipment.
- Working environment in which the equipment is used (e.g. wet or dusty) or the likelihood of mechanical damage.
- Frequency and duration of use.
- Foreseeable abuse of the equipment.
- Effects of any modifications or repairs to the equipment.
- Analysis of previous records of maintenance, including routine visual inspection and combined inspection/testing.

Records of Inspection and Testing

MORE...

HSR25 - *Memorandum of guidance on the Electricity at Work Regulations 1989. Guidance on regulations* available from:

www.hse.gov.uk/pubns/books/hsr25.htm

A suitable log is useful as a management tool for:

- Monitoring the condition of equipment.
- Demonstrating that a system exists.
- Maintaining an inventory of electrical items on site.

Control Measures 10.2

The log should record the equipment tested, date, date of next test, record of faults found and any rectification carried out or removal/replacement of equipment.

Records do not have to be on a paper system, as many PAT meters will record and log data that can be downloaded to a computer. Items should be labelled to show they have been tested ('PASS') and the next inspection/test date.

Advantages and Limitations of In-Service Inspections and Testing (Portable Appliance Testing)

Advantages:

- Satisfies a legal requirement.
- Early recognition of potentially serious equipment faults, such as poor earthing, frayed and damaged cables and cracked plugs.
- Identification of incorrect or inappropriate electrical supply and/or equipment.
- Identification of incorrect fuses being used.
- Reduced number of electrical accidents.
- Misuse of equipment can be monitored.

Limitations:

- Some fixed equipment is tested too often leading to excessive costs.
- Personal equipment, such as phone chargers are never tested since there is no record of them.
- Equipment may be misused or overused between tests.
- Incompetence of the tester.
- Incorrectly calibrated testing equipment.

STUDY QUESTIONS

4. What six protective factors should be used for electrical equipment?
5. What are the three conditions that need to be met to ensure a safe system of work
6. What is the first step in treating the victim of an electric shock?

(Suggested Answers are at the end.)

10.3 Control Measures for Working Underneath or Near Overhead Power Lines

Control Measures for Working Underneath or Near Overhead Power Lines

IN THIS SECTION...

- Precautions are necessary to avoid incidents when working in the vicinity of overhead power lines.
- Such work should be avoided where possible. If it must take place:
  - Ensure management, planning and consultation with service providers to enable supplies to be isolated or diverted.
  - Ensure adequate risk control measures and emergency procedures are in place.
  - Use ground-level barriers, including goal-posts, adequate safety signs and correct clearance distances.

Legal Requirements for Working Near Power Lines

The law requires that work may be carried out in close proximity to live overhead lines only when there is no alternative and only when the risks are acceptable and can be properly controlled.

In particular, Regulation 14 of the **Electricity at Work Regulations 1989** states:

"No person shall be engaged in any work activity on or so near any live conductor (other than one suitably covered with insulating material so as to prevent danger) that danger may arise unless–

- *it is unreasonable in all the circumstances for it to be dead; and*
- *it is reasonable in all the circumstance=s for him to be at work on or near it while it is live; and*
- *suitable precautions (including where necessary the provision of suitable protective equipment) are taken to prevent injury."*

Work may be carried out near power lines only when there is no alternative

Many fatalities and major injuries occur from contact with live overhead power lines. A sensible precaution at all times is to assume all overhead lines are live.

It is not necessary for there to be direct contact with an overhead line - a close approach to them can result in a flashover (arcing) taking place. The higher the line voltage, the greater the risk of this happening.

Work near to or beneath overhead power lines should always be avoided where possible.

Preventing Line Contact Accidents through Management, Planning and Consultation

Where work near or beneath the lines cannot be avoided, contact should be made with the owner (e.g. electricity supplier, railway operator, developer) to discuss the work involved. The purpose of this is to:

- Divert all overhead lines near the work site.
- Make the lines dead (isolate the supply) while work is in progress.
- Schedule the work if they can only be made dead for short periods of time.
- Adopt a safe system of work/procedure/precautions/permit-to-work system.

Control Measures for Working Underneath or Near Overhead Power Lines — 10.3

Three situations generally arise in construction at overhead power lines:

- Work in the vicinity of overhead power lines.
- Plant and equipment will pass under the lines.
- Work will be carried out beneath the lines.

The precautions required in each case are set out below. In all cases, work should be under the supervision of an appointed responsible person.

Use of Barriers to Establish a Safety Zone When Working Near Overhead Lines

When creating safe zones:

- Ground-level barriers are used to prevent close approach (minimum distance six metres unless changed/altered by the owner; it will depend on the line voltage).
- Barriers to be conspicuously marked, (e.g. red/white stripes; plastic flags or hazard bunting).
- Barriers may be stout poles, fences, tension wire fences earthed at both ends, oil drums filled with rubble/concrete, an earth bank (less than one metre), solid timber baulks or concrete blocks.
- Lines of plastic flags or hazard bunting (three to six metres above ground level) may be used, care being taken during erecting to avoid contact with, or approach near to, the live conductors.
- Where mobile plant/equipment is used, the length of the overhanging part of the plant or jib needs to be taken into account.
- Barriers will be required on either one or two sides, depending on access to the worksite.
- No storage is permitted between or under the overhead lines or barriers.

Means of Safely Passing Underneath Overhead Lines

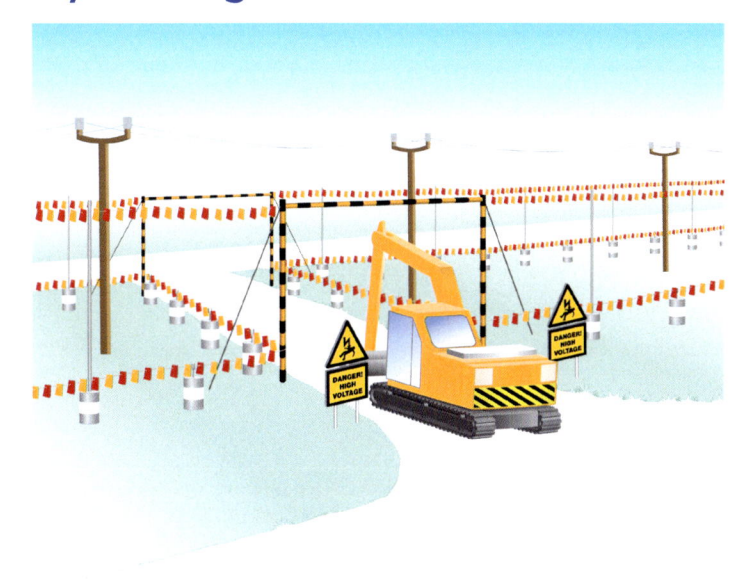

Diagram showing use of barriers, bunting and goal-posts to control proximity of plant to overhead power lines. Based on original source HSG144, The safe use of vehicles on construction sites (2nd ed.), HSE, 2009 (www.hse.gov.uk/pubns/priced/hsg144.pdf)

10.3 Control Measures for Working Underneath or Near Overhead Power Lines

When safely passing underneath:

- A minimum number of defined, fenced, level and well-maintained passageways of restricted width should be used.
- Goal-posts made of timber or plastic pipe should be:
    - Erected at each end of the passageway (parallel to the power lines).
    - Conspicuously marked, e.g. red/white stripes.
- Warning notices must be in place on the approaches to the crossings to:
    - Show the height of the crossbar.
    - Instruct drivers to lower their jibs and keep below this height.
- Electricity proximity warning devices may be fitted on crane jibs, but other safety precautions need to be in place.
- Work after dark requires any notices/crossbars to be adequately and suitably illuminated. Lighting should be at ground level directing the light upwards towards the conductors.
- Extra barriers, goal-posts and warning notices/signs may be required to prevent the upward movement of scaffold poles, crane jibs, excavators and buckets.
- If there is a risk of contact when carrying scaffold poles, ladders or other conducting objects, then they and any mobile equipment should be excluded, or shorter scaffold tubes, ladders or metal sheeting should be used.
- Mobile plant and equipment should be modified by physical restraints to limit their operations, e.g. mechanical stops, limit switches.
- A roof could be constructed over the work area to prevent contact with the live lines.
- The use of proximity warning devices and insulating guards without other safety precautions is unacceptable.
- Care must be taken not to reduce any distances/height clearances during any type of construction work, e.g. dumping, tipping waste, landscaping, scaffolding, etc.
- For work where there are buried services (electricity, gas, water), extra precautions will be required.

Key Emergency Procedures for Contact with an Overhead Line

The HSE's Guidance Note GS6 *Avoiding danger from overhead power lines* recommends a procedure based on the following:

- Never touch the overhead line's wires.
- Assume that the wires are live.
- If lines are dead, they may be switched back on either automatically or remotely if the line operator is unaware they have been struck.
- Call the emergency services.
- Move away as quickly as possible and stay away until the situation has been made safe.
- Either stay in the vehicle or, if you need to get out, jump out of it as far as you can. Do not touch the vehicle while standing on the ground. Do not return to the vehicle until it has been confirmed that it is safe to do so.
- If a live wire is touching the ground the area around it may be live.
- Keep a safe distance away from the wire or anything else it may be touching and keep others away.

Control Measures for Working Underneath or Near Overhead Power Lines | 10.3

STUDY QUESTION

7. (a) What three types of work can be carried out at or near overhead power lines?

   (b) What precautions are required for those three types of work?

(Suggested Answer is at the end.)

Control Measures for Working Near Underground Power Cables

IN THIS SECTION...

- Work must be planned according to the correct legislation.
- Ensure that cable plans; service locators; safe digging practices; and correct tools, locating devices and route planning are used.

Planning the Work

Provision of Pre-Construction Information

This subject, considered in detail in the **Construction (Design and Management) Regulations 2015**, requires the client to provide the principal designer with safety related information that will be of use to contractors carrying out work near underground services. If available to the client, that information would include drawings and plans showing the location of buried services.

Using Cable Plans

Obtaining and Reviewing Plans Before Excavation Work Starts

Before work commences, cable plans or other information should be sought in an attempt to locate the position of all cables in the area in which the excavation is to take place. Where available, the information should be obtained from the owners of the cables, such as the local electricity company, the local authority and private landowners. Plans are frequently inaccurate for a number of reasons, such as repositioning of the cable without the knowledge of the owner, regrading of ground which alters the depth of the cable, removal of the reference points originally used to locate the position of the cable, and snaking of cables in trenches. Above-ground installations such as street lighting, illuminated signs, substations and evidence of backfilled trenches which may house underground cables can also be used.

Most plans will give useful information on the location, number and configuration of cables and will help in the use of cable detectors. However, they cannot be relied upon to give precise information.

What to do if the Information Cannot be Obtained

If information is not available or is thought to be inadequate, extra care must be taken:

- Workers must check the address to ensure they are at the right location.
- Use cable locating devices to establish existing cable runs.
- Mark encroachment lines.
- Do not use power tools within 500mm to marked cable runs.
- Use insulated spades or forks with shortened tines.
- Continue digging by hand until all cables are exposed.

Control Measures for Working Near Underground Power Cables | 10.4

Use of Equipment for Detecting/Locating Buried Services

All personnel who are involved in excavation work where underground cables may be present should be adequately trained in the dangers and the precautions which should be taken. Training should include details on how the hazards occur, types of cables, depths of laying cables, use of plans and location devices, and action to be taken in the event of cable damage.

Use of Service Locating Devices

The location of cables shown on plans should be checked as accurately as possible using a cable locating device. Where no plans have been obtained, the use of the cable detector is additionally important. As cables are located with the device, their routes should be marked on the surface of the ground with waterproof crayon, chalk, paint or wooden pegs as appropriate.

Various cable locating devices are available, the main categories being:

- **Hum detector**: detects the magnetic field radiated by live cables which have a current flowing through them. The disadvantage is that cables with little or no current flowing will not be detected.
- **Radio frequency detector**: receives the low-frequency radio signals which may be picked up and re-emitted by cables. A disadvantage is that objects other than cables may re-radiate the signal and be detected. Results may also vary depending on locality and length of cable. This type of detector is often referred to as a Cable Avoidance Tool (CAT).
- **Transmitter-receiver detector**: a signal is induced into a cable by a transmitter or signal generator which is connected either to the cable or placed close to it. The receiver detects the signal, thereby locating the route of the cable. These devices require more skill to operate and also rely on knowledge of some part of the cable in order that a connection point for the transmitter may be identified. The detector is the same device as mentioned above but used in combination with a signal generator. It is often referred to as a 'CAT and Genny' - cable avoidance tool and signal generator.
- **Metal detector**: this will locate flat metal covers and joint boxes but may not detect round cables. However, it can be useful in the location of possible points of connection for transmitters as above.

The main disadvantage with cable detectors is that they may be unable to distinguish between a number of cables in close proximity and may register them as one cable. If two cables are located one above the other, the upper one may be located but not the lower one. It cannot be assumed that all cables have been successfully identified and therefore use of cable locating devices cannot be relied on in isolation.

Locating devices should only be used by trained operators, within the manufacturer's guidelines, and should be maintained in good working order.

Safe Digging Practices

Once the location of cables has been ascertained using plans and detectors, the routes should be confirmed by digging trial holes which should expose the cables. Hand tools should be used and particular care should be taken above or close to the route of the cable. It is preferable that excavations should be alongside rather than above the cable and that final exposure should be by horizontal digging which is easier to control. Incorrectly used hand tools cause many accidents every year. Accidents may be minimised by the use of spades and shovels rather than other tools such as picks, avoiding spiking or throwing tools into the ground, or by using compressed air tools (such as air-knives) which can expose cables safely. Hand-held power tools and mechanical excavators are a major source of danger and should not be used too close to underground cables.

Certain precautions should be taken once cables have been exposed. In particular, cables:

- With spans of more than 1m should be supported.
- Should not be used as handholds or footholds.
- Should not be moved unless absolutely necessary.
- Should be reinstated with advice from the owner.

10.4 Control Measures for Working Near Underground Power Cables

If cable locations are altered during the work, the cable owner should be notified before backfilling of trenches in order that location plans may be amended.

When backfilling, it is good practice to ensure that cables are buried to the required depth (dependant on voltage) in sand or light soil and with marker tape or tiles buried above the cable. Backfilling with sand or light soil prevents excessive stress and damage to the cable. It also allows for the easy identification of disturbed ground in the event of future excavation work. Buried warning tape or tiles lying above the cable also helps in this respect.

Use of Appropriate Tools, Locating Devices and Route Planning When Undertaking Excavation Work

Route Planning

Once a locating device has been used to determine cable positions and routes, excavation may take place, with trial holes dug using suitable hand tools as necessary to confirm this.

Vacuum Excavation

Vacuum excavation is the process of digging by using high-pressure water or air to soften the soil and then utilising a vacuum to remove the soil and debris. Vacuum excavation is also known as suction excavation or soft dig. Vacuum excavation virtually eliminates the danger of damaging underground utilities. All workers involved must be competent and physically fit to operate the equipment. An appointed support worker will be required to relay information from the machine (vacuum plant) to the boom operator (vacuum head).

Air Excavation

Air or dry excavation is the process of using compressed air to disturb the earth's soil which is then vacuumed up into a debris tank. Air excavation can be slower than hydro excavation, but it does allow the debris to be used back into the hole.

Hydro Excavation

Hydro excavation is the process of removing free-flowing solids, such as aggregates, backfill and material surrounding pipes and sewers. Pressurised water is used to loosen the material, which is then transferred by vacuum into a tanker for removal. This is a more efficient alternative to digging out these materials and then removing them from site in a two-stage operation.

> **STUDY QUESTIONS**
>
> 8. What four cable-locating devices can be used to detect underground cables?
>
> 9. What precautions should be taken once cables have been exposed?
>
> (Suggested Answers are at the end.)

Summary

This element has dealt with some of the hazards and controls relevant to electrical safety in construction.

In particular, this element has:

- Explained the range of effects that electric shock has on the body.
- Described the hazards of electric shock, direct and indirect burns, fire, arcing, short circuits and static electricity.
- Explained the reasons why the selection of suitable work equipment is so important.
- Outlined the dangers of working near power lines, underground cables and on mains electricity supplies.
- Described the various protective systems available - fuses, earthing, isolation, double insulation, reduced low voltage and residual current devices - and their purpose in an electrical system.
- Outlined the importance of using competent people for electrical work.
- Outlined the range of safe systems of work which exist, including those for live working, safe isolation, locating buried services and overhead cables.
- Identified the care that must be taken when approaching and treating an electric shock or burn victim.
- Explained the value of user checks, formal inspection, and combined inspection and testing to ensure the safety of portable electrical equipment and the requirements for record keeping.
- Outlined the precautions needed when working near overhead power lines, passing beneath lines and carrying out work beneath lines.
- Outlined the hazards and precautions when working near underground power cables.

Exam Skills

Question

Scenario

You are in the site office which overlooks part of the site where groundwork has begun. An excavator has started digging an excavation and about 20m along the line of the proposed excavation, overhead power lines cross over this part of the site. You immediately identify the hazard, rush outside and summon the site foreman to stop the work. He looks surprised at your concern and you inform him work cannot commence until appropriate measures are introduced to prevent contact with the overhead power lines. The foreman seems unsure of the risks associated with the power lines and the measures that would need to be put in place.

Task: Electricity

You decide to make a briefing document to highlight how contact could occur whilst working with high-reach equipment on site and how the risk of contact with overhead electricity power lines can be prevented.

(Total: 10 marks)

Approaching the Question

Now think about the steps you would take to answer this question:

Step 1 The first step is to **read the scenario carefully**. Note the question implies two parts, how contact can be made with overhead electricity lines and the second what measures can be introduced to prevent this.

Step 2 Now look at the **task** - prepare some notes under the two parts. You decide to create a briefing document to inform the foreman so you will need to structure your approach using the two headings: how contact can be made and what measures need to be introduced to prevent contact.

Step 3 Next, consider the **marks** available. In this task, there are 10 marks available. You have to create a briefing document that is easy to understand; giving examples for each part can aid understanding. You will need to provide around 9 or more different pieces of information including examples for this task, as some pieces of information may gain more than 1 mark as they will require additional detail. The headings will allow you to keep your response focussed on the two requirements of the question but judgment will have to be made on how much detail is required in each.

Step 4 **Read the scenario and task again** to make sure you understand the requirements and ensure you have a clear understanding of electrical principles. (Re-read your study text if you need to.)

Step 5 The next stage is to **develop a plan** - there are various ways to do this. Remind yourself, first of all, that you need to be thinking about how contact can be made and then how this can be prevented when working under overhead electricity power lines.

Exam Skills

Suggested Answer Outline

How contact with overhead electricity power lines can be made:

- Direct contact by the extendable arm of the excavator.
- Through the effect of arcing – part of the excavator comes within a certain distance of the power line.

Measures to prevent contact with overhead electricity power lines:

- Liaise with the power line supplier to isolate the electric supply or reschedule the work for short periods when isolation of power can be done.
- Creating goal posts under the lines.
- Creating a SSW for the work to be undertaken.
- Putting physical restraints on excavator arms.
- A roof can be constructed over the work area to prevent contact.
- Control of tipping operations via roof construction.
- Work to be supervised at all times.

Now have a go at the question yourself.

Example of How the Question Could be Answered

Working or travelling under overhead electricity power lines can pose significant risk due to the possibility of direct contact with the lines by high-reach equipment such as excavators and cranes. This could occur during extending the booms and whilst slewing the equipment. Another way of coming into contact with electricity is through the effect of arcing which is where, within a certain proximity, electricity may arc (jump) from the conducting surface of the overhead power lines to another conducting surface, the excavator arm or jib of the crane. Both situations could prove fatal to personnel operating the equipment.

In order to mitigate contact with electricity in this setting, the first action would be to contact the electricity supplier to see if the electricity supply can be isolated and therefore eliminate the risk. Alternatively, rearranging the work to enable the supplier to isolate the electricity supply for short windows of time may also be an option. If the overhead electricity supply cannot be isolated then ground level barriers clearly marked are to be used to prevent close approach. Where mobile plant and vehicles have to cross underneath the power lines then the use of goal posts made of timber need to be erected parallel with the power lines. Signs should indicate the height of the crossbar. When working underneath overhead power lines, mobile plant should be modified by physical restraints to limit their operations, e.g. mechanical stops or limit switches. A roof constructed over the work area (made of a non-conducting material, i.e. wood) to prevent contact with the live lines could also be created. This would create a tunnel effect. Care should also be taken not to reduce any distances/height clearances during any type of construction work, e.g. dumping, tipping waste, etc. All these practices should be included in a SSW with supervision to ensure the work practices are followed and no situation is created where contact could occur.

Reasons for Poor Marks Achieved by Exam Candidates

- Not following a structured approach for the briefing document; failing to provide information on the two subject areas of how contact can occur and how contact can be mitigated.
- Not expanding the answer beyond a few words and just using bullet points as opposed to giving a sentence of explanation.
- Misinterpreting the question by focusing on buried services.

Element 11

Fire

Learning Objectives

Once you've read this element, you'll understand how to:

1. Describe the principles of fire initiation, classification and spread and the additional fire risks caused by construction activities in an existing workplace.

2. Outline the principles of fire prevention and the prevention of fire spread in construction workplaces.

3. Identify the appropriate fire detection, fire alarm systems and fire-fighting equipment for construction activities.

Contents

Fire Principles	**11-3**
Basic Principles of Fire	11-3
Classification of Fires and Electrical Fires	11-4
Basic Principles of Heat Transmission and Fire Spread	11-4
Common Causes and Consequences of Fires within the Construction Industry	11-5
Preventing Fire and Spread	**11-8**
Control Measures to Minimise the Risk of Fire Starting in a Construction Workplace	11-8
Fire Alarms and Fire-Fighting	**11-14**
Common Fire Detection and Alarm Systems	11-14
Portable Fire-Fighting Equipment	11-15
Extinguishing Media	11-17
Access for Fire and Rescue Services and Vehicles	11-18
Summary	**11-20**
Exam Skills	**11-21**

Fire Principles

IN THIS SECTION...

- Three things must be present for a fire to start: fuel, oxygen and heat.
- The five classes of fire (determined by the types of fuel) are:
  - Class A (organic solids).
  - Class B (flammable liquids or liquefiable solids).
  - Class C (flammable gases).
  - Class D (metals).
  - Class F (high temperature fats).
- Fire can spread through a workplace by various means of transmission, including direct burning, convection, conduction and radiation.
- Additional risks of fire within construction are faulty or misused electrical equipment, deliberate ignition (arson), hot work, heating and cooking appliances, and smoking.

Basic Principles of Fire

The Fire Triangle

The basic principles of fire and combustion can be represented by the fire triangle.

The three things required for a fire to start are outlined in the image:

The fire triangle

DEFINITION

FIRE

A chemical reaction that uses fuel and oxygen and creates heat and light.

The first thing to come from a fire is invisible vapours. Once the vapours ignite, the fire produces heat, a flame and smoke. Smoke is made up of hot combustion gases such as carbon monoxide, carbon dioxide and solid particles of soot. It will also contain (sometimes toxic) residues from substances that burn (e.g. burning plastics will produce acidic fumes).

The fire triangle is useful for two reasons. It represents:

- **Fire prevention** - keep the three elements apart and fire cannot start.
- **Fire-fighting** - remove one of the elements and a fire will go out.

TOPIC FOCUS

The **fire-fighting** triangle:

- **Remove oxygen** - smother the fire (using a fire blanket, foam, dry powder).
- **Remove heat** - cool with water or carbon dioxide.
- **Remove fuel** - starvation - turn off the gas/electricity/oil supply.

11.1 Fire Principles

Sources of Ignition

An ignition source starts the combustion process. (Once a fire starts, it continues to produce its own heat.) Sources are sparks from hot work such as grinding, cutting and welding, heaters, overloaded electrical equipment, smoking and sunlight.

Fuel and Oxygen in a Construction Workplace

Fuels are flammable or combustible materials or substances, e.g. paper and card associated with packing materials or office stationary, wood from pallets used to transport bricks or timber for structural use, petrol and diesel for site vehicles, propane for gas cylinders used for bitumen heaters, oil, paint and solvents, plastics and foam from sealants.

Oxidising Materials

Oxygen can also come from oxygen-rich chemicals (oxidising agents) such as ammonium nitrate.

Classification of Fires and Electrical Fires

Fires are commonly classified into five categories according to the fuel type. The classification is useful as the basis for identifying which extinguisher to use:

- **Class A** - solid materials, usually organic, e.g. paper, wood, coal, packaging material and textiles.
- **Class B** - flammable liquids or liquefiable solids, e.g. petrol, diesel, paraffin, oil, grease.
- **Class C** - gases, e.g. methane, propane, butane, acetylene and mains gas.
- **Class D** - metals, e.g. aluminium, magnesium.
- **Class F** - high temperature fat, e.g. cooking fat.

Note that there is no Class E fire. This was avoided to prevent confusion between Class E and electricity. Electricity is not a fuel (although it is a common heat/ignition source for unwanted fires).

Basic Principles of Heat Transmission and Fire Spread

Once a fire has started, there are four methods by which it can spread:

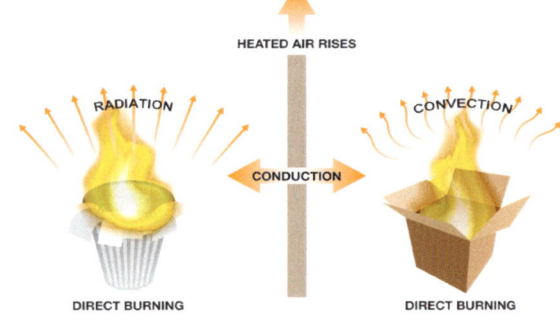

Fire spread

Convection

The principle is that hot air rises and cold air sinks. Hot gases created from the fire rise straight up:

- **In a building**, the gases will hit the ceiling and spread out in a layer beneath it. On contact with other combustible materials, these hot gases may cause them to ignite.
- **Outdoors**, these convection currents carry hot embers, which may fall to the ground and carry the fire to another location.

Conduction

Heat can be transmitted through solid materials. Some materials (metals, and copper in particular) heat very efficiently. Steelwork, pipes and cables running from room to room can carry the heat through and spread a fire.

Radiation

Heat energy can radiate through air as infrared heat waves travelling in straight lines, and can pass through transparent surfaces (e.g. glass). Radiant heat generated from a fire shines onto nearby surfaces and is absorbed, and if the material heats up enough it can burst into flames.

Direct Burning

This is where a flame simply spreads through a material that is on fire, until it is all consumed.

Common Causes and Consequences of Fires within the Construction Industry

There are many reasons why fires start during construction activities. Some of the more common ones are:

- **Hot work** - many construction tasks involve heat and naked flames, e.g. use of propane torches, tar-boilers, asphalting burners, welding and burning. Even some remote site lights may be propane-powered.
- **Electricity** - faulty wiring, especially in temporary site circuits; overloaded conductors; misused equipment; and incorrect use of electrical equipment, especially power tools, in inappropriate environments.
- **Smoking** - especially carelessly discarded smoking materials, such as cigarette ends and matches.
- **Cooking appliances** - fat pans, toasters, etc. in welfare areas left unattended.
- **Heating appliances** - electric fan heaters and portable space heaters, often left unattended and used on site for drying clothes.
- **Deliberate ignition** (arson) - fires started because the construction site is close to public access areas and is an easy target, perhaps by persons not wanting building carried out there, or by young people playing.
- **Unsafe use and storage of flammable substances** - petrol, diesel, propane gas cylinders are used and stored on site.
- **Mechanical heat** - generated by friction between moving parts of site machinery, e.g. motors, bearings, tight drive belts, etc.

Unsafe use and storage of flammable substances can cause fire on site

TOPIC FOCUS

Arson

In criminal law, arson is defined as: *"the act of intentionally or recklessly setting fire to another's property or to one's own property for some improper reason"*.

Arson is a common problem on construction sites for a range of possible reasons, for example:

- The location of the site in relation to other premises, schools, etc. or in an inner city location which makes it vulnerable.
- The site is unsecured, without adequate perimeter fencing.
- The site does not have security officers in attendance or doing patrols.
- There is little or no security lighting.
- Flammable or combustible materials are stored in unsecured areas close to the perimeter of the site.
- Former employees may have a grudge against the contractor or site management.
- Attacks may be made by opportunist vandals, especially children during school holidays.

It is important to consider and control all of these factors to reduce the risk of arson and ensure workers report anyone seen acting suspiciously on or around the site.

11.1 Fire Principles

> **TOPIC FOCUS**
>
> **Friction**
>
> Friction is two or more materials or surfaces moving against one another, giving off heat.
>
> Without lubrication or cooling substance, such surfaces become very hot or produce sparks, either of which may be sufficient to cause ignition.
>
> Friction can:
>
> - Be caused by impact - one material striking another.
> - Result from surfaces rubbing together or 'smearing' (steel, coated with a softer material, is subject to high bearing pressure with sliding or grazing).

Fire can be disruptive on a construction site, particularly because of the many different contractors and their equipment in use and stored on site. Losses are normally covered by insurance.

The biggest danger is to people, which is a greater risk where construction activities are carried out in a public place that remains in use, e.g. the high street. More people die in fires from smoke inhalation than from burns.

Additional Risks of Fire

Some construction methods use different technologies and materials to produce less expensive buildings that are also often quicker to erect than those of conventional materials and building methods. Some of these materials and components, often produced and assembled off-site, are more prone to fire while they are unprotected during construction. This will particularly affect timber-frame buildings.

The consequences of the use of such materials and methods of construction is often faster propagation of flames and much quicker fire spread among partly-assembled or constructed buildings and the parts awaiting assembly into those buildings. This in turn leads to quicker collapse of burning buildings and spread to properties nearby.

Fire Principles | 11.1

STUDY QUESTIONS

1. Identify the process of heat transmission/fire spread shown in the following images.

    (a)

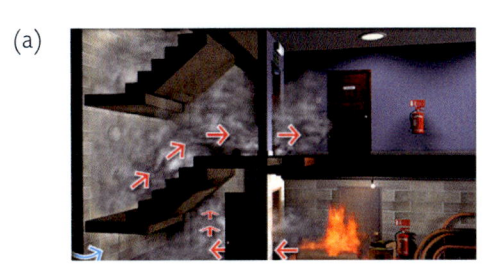

    Source: Safe Practice "Fire Safety"

    (b)

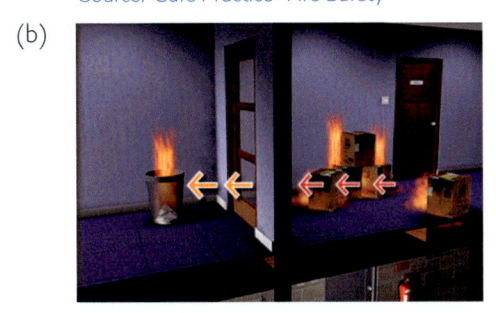

    Source: Safe Practice "Fire Safety"

    (c)

    Source: Safe Practice "Fire Safety"

2. What additional method of heat transfer/fire spread is not illustrated by Question 1?

3. Explain briefly how each of the following might start a fire:

    (a) Friction.

    (b) Space heater.

4. Identify the three ways of extinguishing a fire.

5. Identify the fire classification of each of the following types of fire:

    (a) Butane gas cylinders burning in a storage compound on a construction site.

    (b) Fire in a paint store on a construction site.

    (c) Fire in a construction site office.

(Suggested Answers are at the end.)

11.2 Preventing Fire and Spread

Preventing Fire and Spread

IN THIS SECTION...

- Fire can be prevented by:
  - Eliminating or reducing the use and storage of flammable and combustible materials on site.
  - Correctly storing highly flammable liquids.
  - Controlling ignition sources, including hot work such as welding.
  - Operating safe systems of work, including hot work permits.
  - Maintaining good housekeeping standards.
- Structural measures can help to prevent fire and smoke spread, such as considering the properties of common building materials and the protection of openings and voids.

Control Measures to Minimise the Risk of Fire Starting in a Construction Workplace

Eliminate/Reduce Quantities of Flammable and Combustible Materials Used or Stored

Potential fuels that will be identified in a site fire risk assessment include:

- **Combustible materials** - cardboard packaging, paper, wooden pallets, scaffold boards, etc.
- **Flammable materials**:
  - Liquids such as petrol, diesel, paint and thinners.
  - Gases such as propane, butane and methane.

Control measures (in order of preference):

- **Eliminate** these materials from the site (although total elimination is rarely achievable). Unwanted and old stocks should be removed as a start.
- **Substitute** a high-risk fuel source for a lower risk material, e.g. replacing petrol-driven generators with diesel-powered units, lowering the flammability risk.
- **Minimise** (reduce) the amounts of materials kept and used. This needs:
  - Good stock control and waste management systems.
  - Close control of items such as empty wooden pallets (used for delivering blocks, bags, etc.).
  - A system that ensures regular returns.
- **Safe storage** and **safe use** of materials, for example:
  - Drums of flammable liquid (within the maximum limit) should be on bunded pallets, in a secure fenced compound in a safe location, with fire precautions in place.
  - Separation of combustible and flammable materials into different storage areas.
  - Instead of tipping the drums to get liquid out, use a hand-pump to avoid tipping and spills (and reduce manual handling).

Packaging materials need to be stored safely

TOPIC FOCUS

Storage areas for flammable and combustible materials on site should be:

- Securely fenced, ventilated buildings or open-air compounds.
- Separate from other parts of the site and away from emergency exits.
- Accessible to fire-fighters.
- Properly marked/signed.
- Provided with two escape routes.
- Large enough to allow clear spaces to be maintained around stacks of materials, taking care that the stacked materials themselves do not cause a hazard.

Storage of Highly Flammable Liquids

Liquids (such as petrol and solvents) in industrial products (e.g. paint, ink, adhesives, cleaning fluids) give off flammable vapour. When mixed with air and an ignition source, these vapours can ignite or explode. The flashpoint of a liquid is the minimum temperature that the liquid - under specific test conditions - gives off sufficient flammable vapour to ignite on the application of an ignition source.

The following categories of flammable liquids are:

- Category 3 **flammable liquids** which have a relatively low flash point (between 23°C (73.4°F) and 60°C (140°F) and are therefore relatively easily ignited with an ignition source (such as a match) at normal room temperature.
- Category 2 **highly flammable liquids** which have a lower flash point (<23°C) and a boiling point of 35°C or more and are therefore easier to ignite at normal indoor and outdoor temperatures.
- Category 1 **extremely flammable liquids** which have a similar low flash point (<23°C) and a boiling point of less than 35°C and are therefore very easy to ignite at normal indoor and outdoor temperatures.

When in use in the workplace, the quantity of flammable liquids should be minimised:

- Up to 50 litres of highly flammable liquids.
- Up to 250 litres of flammable liquids.

Storage requirements:

- Store only in appropriate (usually metal) containers with secure lids.
- Containers should be correctly labelled.
- The need to decant highly flammable liquids from one container to another should be minimised, reducing the risk of spillages.
- Area should be well-ventilated.
- Drip trays and proper handling aids should be provided.
- Method/procedure for dealing with spillages and disposal of empty containers and contaminated materials should be in place and understood.

Control of Ignition Sources

Most fires in the workplace are caused by a lack of control over sources of ignition. These are preventable by carefully designed working systems and practices:

- **Electrical equipment** should be routinely inspected and tested to prevent faults that could cause sparks and overheating going unnoticed.
- **Hot work**, such as welding, should be controlled with a permit to work when carried out in sensitive areas.
- **Smoking** should be controlled and limited to restricted areas on site (welfare areas).

11.2 Preventing Fire and Spread

- **Cooking and heating appliances** should be safely located and used carefully. They should not be left unattended.
- **Mechanical heat** (overheating) can be controlled by good maintenance programmes.
- **Bonfires** (often used for burning waste) should not be permitted on site.
- **Deliberate ignition** (arson) should be prevented by good site security, perimeter fences, CCTV, security lighting and good control of combustible waste, i.e. stored away from buildings in a secure area.

Suitable Electrical Equipment in Flammable Atmospheres

The **Dangerous Substances and Explosive Atmospheres Regulations 2002 (DSEAR)** intend to eliminate or reduce the risks of fire (and explosion) arising from the hazardous properties of substances, particularly where they create a flammable or explosive mixture in air.

Part of the elimination or reduction of risks requires the use of intrinsically safe electrical fixtures, fittings and equipment in such atmospheres. Where such atmospheres may occur, the area must be split into hazardous and non-hazardous zones and marked with signs at points of entry.

The zones and categories of electrical equipment are shown below.

Zone 0	Zone 1	Zone 2
Explosive gas atmosphere present continuously or for long periods.	Explosive gas atmosphere likely to occur in normal operation.	Explosive gas atmosphere not likely to occur in normal operation; if it does, it is only for short time.
Cat 1 Electrical Equipment	**Cat 2 Electrical Equipment**	**Cat 3 Electrical Equipment**
'i' intrinsically safe. **BS EN 60079-11:2012**	'd' flameproof enclosure. **BS EN 60079-1:2014**	Electrical type 'e'. **BS EN 60079-7:2015 Explosive atmospheres. Equipment protection by increased safety 'e'.**

In addition, suitable earthing of all equipment and plant is required, and all maintenance controlled to prevent sparks being created (use of a permit to work).

Use of Safe Systems of Work

These should be in place to minimise the risk of fire. Risk assessment will determine what systems are appropriate, including:

- **Permits to work** - controlling hot work, work on electrical systems.
- **Careful control of storage areas** - especially where flammable materials are kept.
- **Refuelling** - careful refuelling of vehicles and plant on site in safe areas.
- **Prevention and control of spillages** - use bunded storage systems and pump-transfer (instead of pouring from drums).
- **Control of waste** - segregation in areas away from fire risks and regular removal of waste materials from the site.

Good Housekeeping

Good housekeeping is fundamental to fire safety and should ensure that:

- Combustible and flammable materials are regularly removed from work areas.
- Items that can't be removed are covered with fire-retardant blankets.
- Waste bins are emptied regularly, so that there is no accumulation of combustible materials.
- Site areas are regularly cleaned and kept free of litter and rubbish.

- Safe disposal of all waste materials is arranged. 'Unofficial' rubbish burning must be banned.
- Skips are placed at least three metres from buildings and other structures.
- Pedestrian routes are always kept clear.

Structural Measures to Prevent the Spread of Fire and Smoke

Building design can be a significant factor in preventing both the outbreak and spread of fire. The main features which influence this are the:

- Layout and construction of the building or the site premises.
- Materials with which buildings are constructed and those used in decoration and furnishings.

Properties of Common Building Materials

Fire affects different building materials in different ways. Controls must therefore ensure that appropriate materials are used in a structure.

Material	Fire Characteristics
Concrete	Usually very resistant to fire and does not collapse catastrophically. May 'spall' - throw off small chunks.
Steel	Severely affected by high temperature - will expand, twist and warp (buckle). Structural elements can be pushed apart, leading to catastrophic collapse. It will also conduct heat and increase the possibility of fire spread.
Brick	Fired clay bricks are very resistant to fire.
Timber	Thin timber will burn easily and quickly fail. Thicker timber may char on the outside, protecting its inner core - it will fail slowly.

To strengthen construction materials in a fire situation:

- Concrete is mixed with lighter aerated aggregates and reinforced with steel. Fire resistance depends on the type of aggregate used and the thickness of the concrete over the reinforcing rods.
- Steel is usually strengthened by encasing it in concrete or fire-resistant board.
- The integrity of bricks is enhanced by thickness, rendering or plaster, being a non-load-bearing wall, and the lack of perforations or cavities in the bricks.
- Timber and other surfaces can be treated with fire retardant or resistant finishes and intumescent coatings.

TOPIC FOCUS

The **fire resistance of timber** depends on the 'Four Ts':

- The **thickness** or cross-sectional area of the piece.
- The **tightness** of any joints involved - in general, the fewer joints, the better.
- The **type** of wood - generally, denser timber has better resistance (the surface chars, but because conduction is poor, the internal timber still performs structurally).
- Any **treatment** received, e.g. flame-retardant treatment is often applied to such materials as plywood or chipboard sheets.

11.2 Preventing Fire and Spread

Other Building Materials

Other building material include:

- **Stone** - granite, limestone and sandstone are commonly used for cladding. It provides good fire resistance but may be subject to 'spalling' when used as cladding.
- **Building blocks** - clay or concrete, used as bricks in structures - provide excellent fire resistance, provided that the foundations and supporting structure can keep the wall in place during the fire.
- **Building boards** - used for cladding and insulation; are combustible but not easily ignited.
- **Building slabs** - thick versions of board, used as substrates for roofing. Fire resistance is dependent on the materials used such as plywood, OSB (a structural panel made of wood strands sliced in the long direction and bonded together with a binder under heat and pressure) or timber boards; although concrete, wood wool or profiled metal can also be used. Current building regulations require all building slabs to be fire resistant, with more stringent standards for those used in buildings over 18m high.
- **Glass** - will break and cannot be used as a fire barrier, unless wired or treated (copperlight).
- **Insulating materials** - modern materials are non-combustible.
- **Lime** (plaster) - generally has good fire resistance.
- **Paint** - most paints are flammable; combustible when a dried finish. Flame-retardant and intumescent paints are available. When exposed to heat, these bubble rather than burn, giving additional protection to the painted timber.
- **Plastics** - thermoplastics will easily melt; thermosetting plastics will not, but will deteriorate in a fire.
- **Fire doors** - to allow movement between compartments, fire doors are fitted. These should:
    - Be rated to withstand fire for a minimum of 30 minutes.
    - Have a self-closing device.
    - Be fitted with an intumescent strip.
    - Have a vision panel of fire-resistant glass.
    - Be clearly labelled (e.g. 'Fire Door - Keep Shut').

DEFINITION

INTUMESCENT STRIP

'Intumescent' means 'swelling up when heated'.

Intumescent strips, seals or foam are placed in door frames and other openings (e.g. service ducts between floors) and will swell with the heat from a fire and seal the gap, preventing the passage of air (which might feed the fire), smoke and flames.

Compartmentation

If a fire starts within a building, it should be contained and prevented from spreading. This can be done by designing the building in such a way that it is divided up into **compartments**, each surrounded by fire-resistant materials.

There are two types of fire compartment or cell:

- Those designed to keep a fire in - areas of high risk such as plant rooms or flammable stores.
- Those designed to keep a fire out - high-loss-effect areas such as document archives or computer rooms, and escape routes.

The walls, floors and doors which form the boundary to a fire compartment must generally provide a 60-minute resistance to fire.

The walls, floors and doors which subdivide fire compartments must generally provide a 30-minute resistance.

Protection of Openings and Voids

Voids beneath floors and above ceilings, as well as openings around pipework and other services, lift shafts, air handling ducts, etc. can allow air to feed a fire, as well as assisting in the spread of fire and smoke. To help control this:

- Debris should not be allowed to accumulate in voids.
- When necessary, openings should be bonded or fire-stopped with non-combustible material.
- Ventilation ducts and gaps around doors must have the facility to be stopped in the event of a fire. This can be achieved by the use of baffles, self-closing doors and intumescent material which expands when subject to heat, thereby sealing the opening.
- It is important that any new openings made in a fire-resistant compartment are reinstated or protected in some way, e.g. when cables are run through a hole in a wall, the opening could be filled with a spray-in intumescent foam.

DEFINITION

FIRE STOPPING

A **firestop** is a fire protection system made of various components used to seal openings and joints in fire-resistance rated wall or floor assemblies. For penetrating cables, these are also known as Multi Cable Transits (MCTs).

STUDY QUESTIONS

6. How might you minimise the risk of fire in a woodworking area?
7. What precautions should be taken when using flammable liquids?
8. How are flammable liquids categorised?
9. How can good housekeeping on construction sites reduce the risk of fire?
10. Upon what does the fire resistance of each of the following building materials depend?
    (a) Timber.
    (b) Reinforced concrete.
    (c) Brick walls.
11. Describe the effects of fire on an unprotected steel beam.
12. Describe how flame-retardant paint protects covered timber.
13. Describe the conditions that determine the three zones used to identify hazard areas (**DSEAR**).

(Suggested Answers are at the end.)

11.3 Fire Alarms and Fire-Fighting

Fire Alarms and Fire-Fighting

IN THIS SECTION...

- Fire safety relies upon systems to detect and warn of fire, and a means of fighting a fire.
- Consideration of fire safety systems must include:
  - Common fire detection and fire alarm systems.
  - The siting, maintenance, inspection and training requirements with regard to portable fire-fighting equipment.
  - The advantages and limitations of extinguishing media: water, foam, dry powder, carbon dioxide, wet chemical and specialist powder.

Common Fire Detection and Alarm Systems

Fire Detection

In the simplest of workplaces, where all parts can be seen by occupants and the risk of fire is low, fire detection can rely on nothing more than a person seeing or smelling the fire. In larger workplaces, or where early detection is critical, e.g. remote or unoccupied places, detection systems should be used:

System	Features
Smoke/fume detectors	Very common. Will detect small particles in smoke, are very sensitive and give early warning. They have ionising or optical sensors, but can give false alarms in humid, dusty or smoky atmospheres.
Heat detectors	More suitable for some situations. Detect heat from a fire, but are less sensitive and give later warnings. They can detect heat by fixed temperature or the rate of rise in temperature (fusion or expansion heat detectors). They may not detect a slow, smouldering fire giving off smoke but little heat.
Flame detectors	These are optical sensors and will detect flames by ultraviolet and infrared systems.

Fire Alarms

On small, compact sites, word of mouth might be adequate, but in all other cases, alarm systems should be fitted as early as possible in the construction phase, maintained in good working order and repaired.

Problems can arise if wiring operations lead to the alarm system being disabled (deliberately or otherwise) even for a short length of time, particularly where construction occurs in an existing workplace. Alternative means of raising the alarm need to be planned in the event of this happening, and the occupants informed.

All alarm systems must be maintained and tested regularly and the results recorded. Any faults discovered must be rectified and the system re-checked.

Fire alarm

Manual Systems

These are suitable for small workplaces of low risk and include rotary gongs, iron triangles, hand bells, whistles and horns. They can only raise the alarm over a limited area and for a limited time. There should be a means for the person raising the alarm to make it more widespread - by using a phone or public address system, or a manual/electric system.

Interlinked Smoke Alarms

Used in remote locations, these units detect the smoke (or flames) and sound an alarm, not just stand-alone, but often in a linked circuit.

Manual/Electric Systems

These are systems which are initiated from an alarm call point (a 'break glass' unit). When pressed, the alarm is sounded throughout the premises (or a particular part of them). It may also relay an alert to the fire service.

Automatic Fire Alarms

These are not normally present on small construction sites. They are made up of automatic detectors and manual call points linked through a central control box to alarms (and sometimes flashing beacons). A person can activate them on seeing a fire, or they will initiate automatically if no-one is around.

Note: In noisy environments, visible warnings (flashing lights) may be needed in addition to audible alarms. Similar provision may be appropriate for hearing-impaired workers.

Portable Fire-Fighting Equipment

Small fires can be dealt with quickly using portable fire extinguishers, but this should only be attempted where the person is trained to do so and it is safe, without putting them at risk.

The following extinguishing media are used:

- **Fire extinguishers** - coloured red, with an identifying colour code to denote the extinguishing agent contained (e.g. water, carbon dioxide, foam, powder or vaporising liquid).
- **Fire blankets** - a fibre blanket used to smother small fires. Very useful in a kitchen where there may be burning fat, and also for smothering burning clothing.
- **Hose reels** - sited in fixed locations in buildings to allow fire teams to fight larger fires.
- **Sprinkler systems** - built-in systems sited in buildings, warehouses and at large flammable tank locations - they work automatically off (usually) thermal detectors.

Portable fire extinguisher

Note: Where sprinkler systems are installed, they should be deactivated before hot work is carried out that could set them off.

11.3 Fire Alarms and Fire-Fighting

> **TOPIC FOCUS**
>
> ### Fire Extinguishers
>
> **Under BS EN 3-10:2009**, all fire extinguishers are now **red**, with colour identification to denote the extinguishing agent contained.
>
> Common types of fire extinguishers and their uses include:
>
> - **Water** (red with white lettering) - suitable for Class A fires. It works by cooling the fire. A standard water extinguisher is not suitable for use on Class B, D or F fires or live electrical equipment (this might lead to risk of shock). Certain specialised water extinguishers are available for use on Class B and F fires.
> - **Carbon dioxide** (black colour coding) - suitable for Class B fires and fires involving live electrical equipment. It works by displacing oxygen. It is not suitable for use on Class D or F fires. It must be used with care because the body of the extinguisher gets very cold during use and can cause a freeze-burn injury. Carbon dioxide is an asphyxiant gas and so care must be exercised when using in an enclosed space.
> - **Foam** (cream colour code) - suitable for Class A and B fires. It works by smothering the fire or by preventing combustible vapours from mixing with air. Some specialist foam can be used on electrical fires but, again, you must be certain that you are using the right type. As the foam is wet, it is not suitable for Class F hot fat fires.
> - **Dry powder** (blue colour coding) - suitable for all classes, with the exception of Class F, and use on live electrical equipment. It works by cooling the flames and may chemically interfere with the combustion process. It can be very messy and the powder must not be inhaled.
> - **Dry powder** (violet colour coding) - suitable for Class D.
> - **Halon** (green colour code) - suitable for Class A and B fires, especially live electrical equipment. However, under the Montreal Protocol halon is now very limited and is replaced with other gas or vaporising liquids.
> - **Fire blankets** (in red box or wrapper) - suitable for Class B and F fires.

Siting

Portable fire-fighting equipment should be:

- Located at fire exit doors and along escape routes.
- Positioned close to specific hazards (e.g. a flammable store) on site, in clearly marked locations.
- Easily accessible (kept clear) at all times.

Maintenance and Training Requirements

To ensure they are always available and work when we need them, extinguishers are to be routinely inspected and maintained by means of:

- **Frequent routine inspections** - to make sure they are in their correct locations and that they appear to be full and in working order (with the firing pin still tagged, or a gauge reading full). This can be done as a housekeeping check or routine site inspection.
- **Planned preventive maintenance** - a regular service usually done by qualified persons or fire safety engineering companies. This ensures the condition of the extinguisher body, its operating system and its contents.

Employees need enough fire safety training to:

- Understand the theories and principles of fire.
- Be able to raise the alarm when necessary.

Fire Alarms and Fire-Fighting — 11.3

- Know what actions to take if they hear the alarm.
- Know when and when NOT to tackle a fire.
- Know when to leave a fire that has not been extinguished.

> **MORE...**
>
> **Fire Safety (Employees' Capabilities) (England) Regulations 2010**
>
> www.legislation.gov.uk/uksi/2010/471/contents/made
>
> **Fire Safety (Employees' Capabilities) (Wales) Regulations 2012**
>
> www.legislation.gov.uk/wsi/2012/1085/contents/made

Training on the use of fire extinguishers should be enough so that they can:

- Identify the correct type to use on a particular class of fire (and which not to use).
- Use a fire extinguisher to effect their escape or save their life in a fire situation.

The intention is not to make them fire-fighters, and employers must take into account the health and safety capabilities of employees when entrusting them with fire safety tasks. This will apply at all levels of employee training, including competent persons, fire marshals, etc.

Extinguishing Media

All extinguishers have the limitation of short duration and relatively small amount of extinguishing agent; therefore they are only suitable for small fires. Some other advantages and limitations include:

Extinguishing Agent	Advantages	Limitations
Water	Good cooling medium for Class A fires. No chemicals involved. Inexpensive material.	Not suitable on most Class B, D or F fires or on live electrical equipment.
Foam	Suitable for Class B fires and fires involving live electrical equipment (if tested to 35kV). Smothers a fire. Valuable where burning liquids are 'running' (moving along the ground, as in a spillage).	Not suitable for Class C, D or F fires. Messy. Not easy to correctly use unless trained.
Dry powder	Suitable for Class A, B, C and D fires, and on live electrical equipment. Smothers a fire.	Not suitable for Class D or F fires. Can be messy. Some noise when exhausting the powder. Not suitable for indoors

11.3 Fire Alarms and Fire-Fighting

Extinguishing Agent	Advantages	Limitations
Carbon dioxide	Smothers quickly. Non-toxic. Suitable for Class A and B fires and live electrical equipment.	A gas cylinder under pressure. Not suitable on Class D fires. Use with care - rapidly exhausting gas can cause freeze-injury if touched. Noisy and can startle a user.
Wet Chemical	Best fire medium for fire involving cooking fats. Has a Class A rating and a Class F rating. Lance means the operator is always a safe distance from burning oils or fats. No dry chemical to clean up.	Expensive to buy and refill. Limited application - must not be used on Class D fires.
Specialist Powder	Designed specifically for use on Class D fires	Can be messy. Some noise when exhausting the powder.

> **TOPIC FOCUS**
>
> **AQUEOUS FILM FORMING FOAM (AFFF)**
>
> Aqueous Film Forming Foam (AFFF) extinguishers have a dual Class A and Class B rating which allows them to be used against both solid and liquid burning fires. These extinguishers also have a conductivity rating of 35kV which means that, although they are not specifically designed for use on electrical fires, they can be safely used on electrical equipment up to 1,000V.
>
> The reason why an "aqueous" medium can be used on electrical equipment is that the method of delivery is by spray nozzle, which breaks up the flow of extinguishant. This prevents a continuous electrical path between the user and the electrical apparatus.

Access for Fire and Rescue Services and Vehicles

The Association of Chief Fire Officers has clarified these requirements as follows:

- A 3.7 m carriageway (kerb to kerb) is required for operating space at the scene of a fire. Simply to reach a fire, the access route could be reduced to 2.75 m over short distances, provided the pump appliance can get to within 45 m of dwelling entrances.
- Access routes must always be kept clear and unobstructed. Any site vehicles parked inconsiderately, even for a short time should be removed immediately. Other obstructions like stacked materials must not be allowed to develop.

Fire Alarms and Fire-Fighting 11.3

STUDY QUESTIONS

14. What are the limitations of manual alarm systems and how may they be overcome?

15. Identify the three ways in which fire may be detected and state the types of automatic detector associated with each.

16. Identify the classes of fire for which each of the following extinguishing agents/devices are suitable:

    (a) Water.
    (b) Carbon dioxide gas.
    (c) Dry powder.
    (d) Foam.
    (e) Fire blankets.

17. State the colour coding requirements for portable fire extinguishers.

18. Outline the main points to be covered in training in the use of fire extinguishers.

(Suggested Answers are at the end.)

Summary

This element has dealt with hazards and controls relevant to fire in the workplace.

In particular, the element has:

- Outlined basic principles of fire safety - including the fire triangle, the five classes of fire, how fire can spread, and some common causes of fire on construction sites.
- Explained how fire can be prevented by controlling potential fuel sources (e.g. safe storage and use of flammable liquids) and potential sources of ignition (e.g. hot work), including the use of suitable electrical equipment in flammable atmospheres.
- Outlined the structural measures that assist in containing fire and smoke if a fire starts, and how to protect openings and voids.
- Described the general principles of fire detection and alarm systems.
- Discussed the main types of fire extinguishers, such as water, carbon dioxide, foam and dry powder, and the advantages and limitations of each type.
- Considered the main requirements relating to access for fire and rescue services and vehicles.

Exam Skills

Question

Scenario

You are in a meeting with the construction project management and during the meeting a concern has been raised by fire and, in particular, the concerns over arson attacks. There is some scepticism amongst some managers about the risks of arson attacks and how to manage against preventing them.

Task: Fire

You are asked by the site manager to give an overview of the risks of arson and, in particular:

(a) Reasons why some construction sites may be vulnerable to arson attacks. **(5 marks)**

(b) How can the risk of arson be reduced on a construction site **(5 marks)**

(Total: 10 marks)

Approaching the Question

Now think about the steps you would take to answer this question:

Step 1 The first step is to **read the scenario carefully**. Note the question is focussing on a particular aspect of fire – arson, so think about 'who' (deliberate act by a person) and 'how' (think fuel and ignition sources) an arson attack could occur.

Step 2 Now look at the **task** - prepare some notes under the two headings "Reasons for arson attacks" and "Ways to reduce arson attacks".

Step 3 Next, consider the **marks** available. In this task, there are 5 marks available for the first part and 5 marks for the second part of the question. Tasks that are multi-part are often easier to answer because there are additional signposts in the question to keep you on track. In this task, you have to create a briefing document that is easy to understand, giving examples for each part can aid understanding. You will need to provide around 10 or more different pieces of information including examples for this task. The headings will allow you to keep your response separate – this will also help the examiner when marking.

Step 4 **Read the scenario and task again** to make sure you understand the requirements and ensure you have a clear understanding of the term 'arson' and how it can be reduced. (Re-read your study text if you need to.)

Step 5 The next stage is to **develop a plan** - there are various ways to do this. Creating a bullet point list could be one way.

Exam Skills

Suggested Answer Outline

Reasons for arson attacks:

- Location of the site (inner city location).
- Lack of site security.
- No security lighting.
- Flammable products stored in unsecured areas.
- Former employees have a grudge.

Measures to reduce the risks of arson attacks:

- Ensure a secure perimeter fence.
- Provide external security lighting and cameras.
- Control all combustible waste.
- Control all flammable products.
- Encourage staff to report incidents of people acting suspiciously.

Now have a go at the question yourself.

Example of How the Question Could be Answered

The reasons why arson can occur on construction sites can be its location such as next to other premises such as schools or in inner city locations where it can attract attention due to the work being conducted. Another reason is through lack of site security which might be both physical - security guards or lack of boundary fencing. Arson attacks normally occur out of working hours under the cover of darkness so insufficient lighting would assist arsonists. Fuel sources for arson often involve access to on-site materials such as waste or flammable products which are not secured. Finally, the arsonist might be an ex-employee who left employment under less favourable terms and therefore bears a grudge.

In order to reduce the risk of arson, site security must be the first consideration by ensuring the site is secure at all times. This can be achieved by installing well maintained security fencing and with dedicated access and egress points which are controlled during normal working hours and securely locked during silent hours and the use of CCTV cameras which can monitor and record activities on site at all times. In addition, ensuring the site is well illuminated, especially in locations where an arson attempt could be made. Ensuring all waste is secured in lockable waste storage receptacles and emptied on a regular basis to minimise the amount at any time. The storage of flammable materials should be kept in a fireproof storage medium, under lock and key and with controlled access. Finally, encourage staff to report incidents of people acting suspiciously on or around the site.

Reasons for Poor Marks Achieved by Exam Candidates

- Not following a structured approach for the briefing document; failing to provide information on the two subject areas.
- Not expanding the answer beyond a few words as opposed to giving a sentence of explanation.
- Misinterpreting the question by focusing on general ignition and fuel sources for a fire.

Element 12

Chemical and Biological Agents

Learning Objectives

Once you've read this element, you'll understand how to:

1. Outline the forms of, classification of, and the health risks from exposure to, hazardous substances.

2. Explain the factors to be considered when undertaking an assessment of the health risks from substances encountered in construction workplaces.

3. Explain the use and limitations of occupational exposure limits, including the purpose of long-term and short-term exposure limits.

4. Outline control measures that should be used to reduce the risk of ill health from exposure to hazardous substances.

5. Understand the prevalence of occupational lung disease among construction workers.

6. Outline the hazards, risks and controls associated with specific agents.

7. Outline the basic requirements related to the safe handling and storage of asbestos on construction sites.

Contents

Hazardous Substances — 12-3

Introduction to Forms, Classification and the Health Risks from Hazardous Substances — 12-3
Forms of Chemical Agent — 12-3
Forms of Biological Agents — 12-4
Health Hazards Classifications — 12-4

Assessment of Health Risks — 12-7

Routes of Entry — 12-7
Factors that Need to be Taken into Account When Assessing Health Risks — 12-10
Sources of Information — 12-11
Limitations of Information Used When Assessing Risks to Health — 12-13
Role and Limitations of Hazardous Substance Monitoring — 12-14
Purpose of Occupational Exposure Limits and How They Are Used — 12-19

Control Measures — 12-22

The Need to Prevent Exposure — 12-22
Adequately Control Exposure — 12-22
Principles of Good Practice — 12-23
Common Measures Used to Implement the Principles of Good Practice — 12-23
Additional Controls for Carcinogens, Asthmagens and Mutagens — 12-33

Specific Agents — 12-35

The Prevalence of Occupational Lung Disease Among Construction Workers — 12-35
Health Risks, Controls and Likely Workplace Activities/Locations Where They Can be Found — 12-36
Health Risks from and Controls for Working with Asbestos — 12-40
Duty to Manage Asbestos — 12-41

Summary — 12-48

Exam Skills — 12-49

Hazardous Substances

IN THIS SECTION...

- In construction activities, many different forms of chemical hazards occur - dusts, fibres, fumes, gases, mists, vapours and liquids.
- Biological agents, such as fungi, bacteria and viruses, can be hazardous to health.
- Chemicals are classified according to their hazardous properties: toxic, harmful, corrosive, irritant or carcinogenic.
- There are differences between acute and chronic health effects of hazardous substances.

Introduction to Forms, Classification and the Health Risks from Hazardous Substances

Exposure to chemical hazards can occur:

- Intentionally - by using chemicals in our work.
- Unintentionally - from spillages and accidents.

In either case, exposure has to be prevented, and where we can't prevent it, it must be controlled so that no harm is caused to those who may be exposed.

Exposure can lead to immediate health effects (e.g. carbon monoxide can cause asphyxiation) or even physical effects (battery acid can burn the skin).

Some hazardous substances can have both short-term and long-term effects, e.g. concrete or stone grinding dust can cause immediate coughing and respiratory distress, and can lead on to permanent lung damage from prolonged or repeated exposure.

Spillages can lead to unintentional exposure to hazardous substances

Forms of Chemical Agent

Chemicals may be in the form of a substance (a chemical element or compound) or a preparation (a mixture of substances). These exist in a variety of physical states and this will affect the way chemical hazards occur in construction activities. The physical forms of chemicals are:

- **Dusts** - small solid particles created by grinding, polishing, blasting, road sweeping and mixing materials (e.g. cement), which become airborne.
- **Fibres** - asbestos and other Machine-Made (formerly Man-Made) Mineral Fibres (MMMF) have different characteristics from dust particles. Important dimensions are the length and diameter of the fibre and the length to diameter ratio.
- **Fumes** - fine solid particles which are created by condensation from a vapour (e.g. welding fume) given off in a cloud. Metallic fume is usually the oxide of the metal and is toxic.
- **Gases** - a formless chemical which occupies the space in which it is enclosed (e.g. carbon dioxide, acetylene).
- **Mists** - small liquid droplets (aerosol) suspended in the air, created by activities such as paint spraying.
- **Vapours** - the gaseous form of a liquid or solid substance at normal temperature and pressure (e.g. solvent vapours given off by acetone).
- **Liquids** - a basic state of matter; free flowing fluid (e.g. water at room temperature).

The form they are in can significantly affect how they might enter the body (discussed later).

12.1 Hazardous Substances

Forms of Biological Agents

Biological agents are micro-organisms. We will look at three types:

- **Fungi** - plant matter lacking chlorophyll and reproducing by spores. Examples include mushrooms, mould and yeasts. Fungal diseases can appear as asthmatic and/or influenza-type symptoms from inhaling dust or air contaminated by fungi, such as dry rot in roofs, or fungal infections such as athlete's foot.
- **Bacteria** - single-cell organisms found in vast numbers in and on the human body. Some are harmless, some are beneficial (certain gut bacteria) and some cause diseases, e.g. Legionnaires' disease or Weil's disease (leptospirosis). Construction activities near waterways could pose a risk from Weil's disease.
- **Viruses** - very small infectious organisms that increase by hijacking living cells to reproduce and generate more viruses. Many cause disease, e.g. hepatitis and AIDS.

Health Hazards Classifications

The **Globally Harmonised System of Classification and Labelling of Chemicals (GHS)** is a single internationally agreed system of chemical classification and hazard communication using labelling and Safety Data Sheets (SDS).

It includes harmonised criteria for the classification of:

- **Physical hazards**, e.g. explosive, oxidising, highly flammable.
- **Health hazards**, e.g. acute toxicity, corrosive, health hazard, serious health hazard.
- **Environmental hazards**, e.g. harmful to aquatic organisms, dangerous for the ozone layer.

> **TOPIC FOCUS**
>
> The main **classifications of chemicals hazardous to health** can be summarised as follows:
>
> - **Acute toxicity** (or very toxic) - small quantities cause death or serious ill health if inhaled, swallowed or absorbed via the skin.
> - **Health hazard** - an immediate skin, eye or respiratory tract irritant, or narcotic.
> - **Corrosive** - destroys living tissue on contact, such as sulphuric acid and hydrochloric acid in chemical cleaners, e.g. for masonry or brickwork.
> - **Serious health hazard** - a cancer-causing agent (carcinogen) or substance with respiratory, reproductive or organ toxicity that causes damage over time (a chronic or long-term health hazard).

Criteria for classifying chemicals have been developed for the following GHS health hazard classes:

- **Acute Toxicity**

    These chemicals cause acute toxic effects after ingestion, skin absorption or inhalation. They are allocated to one of five toxicity categories and category 1 toxic chemicals are those requiring the lowest dose to cause a toxic response.

- **Skin Corrosion/Irritation**

    These chemicals cause:

    - irreversible corrosive damage to the skin; or
    - irritation of the skin which is reversible.

- **Serious Eye Damage/Eye Irritation**

    These chemicals cause:

    - serious tissue damage in the eye or serious physical decay of vision; or
    - irritation of the eye which is reversible.

Hazardous Substances | 12.1

- **Respiratory or Skin Sensitisation**

    These chemicals cause sensitisation, which means they can produce an allergic reaction that will gradually worsen as exposure is repeated. There are two types:

    - **Respiratory sensitisers** - these can cause asthma and similar effects if inhaled (e.g. s and isocyanates).
    - **Skin sensitisers** - these can cause allergic dermatitis on contact with the skin (e.g. epoxy resin used in adhesives and paints). Bad cases can cause absence from work. It can be reportable under **RIDDOR** in certain cases.

DEFINITION

DERMATITIS

A skin disease (sometimes called 'eczema') in which the skin's surface protective layer is damaged, leading to redness/swelling of hands and fingers, cracking of skin and blisters on hands/fingers, flaking/scaling of skin, and itching of hands/fingers with cracks.

MORE...

Information on dermatitis is available in the HSE publication INDG233 *Preventing contact dermatitis and urticaria at work* available at:

www.hse.gov.uk/pubns/indg233.pdf

- **Germ Cell Mutagenicity**

    Mutagenic chemicals may cause genetic mutations that can be inherited.

- **Carcinogenicity**

    Carcinogenic chemicals may induce cancer or increase its incidence.

- **Reproductive Toxicity**

    Chemicals that are toxic to reproduction may cause sterility or affect an unborn child. Known as 'teratogens', they are substances that cause harm to the foetus or embryo during pregnancy, causing birth defects while the mother shows no signs of toxicity. Common teratogens include ethanol, mercury compounds, lead compounds, phenol, carbon disulfide, toluene and xylene.

- **Specific Target Organ Toxicity (Single and Repeated Exposure)**

    All significant health effects, not otherwise specifically included in the GHS, that can impair function (both reversible and irreversible, immediate and/or delayed) are included in this class. Narcotic effects and respiratory tract irritation are examples of this.

- **Aspiration Hazard**

    Aspiration is the entry of a liquid or solid directly through the mouth or nose, or indirectly from vomiting, into the trachea and lower respiratory system. Some hydrocarbons (petroleum distillates) and certain chlorinated hydrocarbons are aspiration hazards. Acute effects include pneumonia, varying degrees of pulmonary injury or death.

Acute and Chronic Health Effects

It is important to understand the difference between acute (short term) and chronic (long term) health effects from exposure to hazardous substances:

- **Acute effects** occur quickly after exposure (i.e. in seconds, minutes or hours), often from large amounts of a substance, e.g. inhaling high concentrations of chlorine gas causes immediate respiratory irritation. These effects are often reversible.

12.1 Hazardous Substances

- **Chronic effects** take time to appear (i.e. months or even years), after exposure to smaller amounts of a substance over a longer period of time, e.g. working with lead can take months to accumulate high levels of lead in the blood. These effects are mostly irreversible.

In terms of prevention, chronic effects present the most difficult control problems. This is because:

- The effects occur over a long period, so the hazard is not recognised.
- The level of contamination required to produce chronic effects is often tolerated by people because they do not experience acute symptoms.
- Symptoms occur slowly, so they are not recognised until an advanced condition of harm has developed.
- When symptoms are recognised, the harm may be too advanced for full recovery - sometimes no recovery is possible.
- Symptoms are often confused with 'normal' ill health or with 'getting older'.
- Symptoms are not always easily identifiable in groups of people with the same exposure, owing to the effect of differing 'personal' metabolisms.

Many hazardous substances can have an acute **and** chronic effect. For example, inhaling solvent vapours can have an almost immediate narcotic effect (acute) and long-term repeated exposure to lower levels can cause liver damage over a number of years (chronic).

STUDY QUESTIONS

1. State the physical forms of chemical agents which may exist in the workplace.
2. Identify the five main health hazard classifications of chemicals.
3. Define the characteristics of mist and fumes, and identify a potential source of each in construction activities.
4. Distinguish briefly between acute and chronic ill-health effects.

(Suggested Answers are at the end.)

Assessment of Health Risks

IN THIS SECTION...

- There are four main 'routes' by which hazardous substances enter the body: inhalation, ingestion, absorption through the skin and injection through the skin.
- The body has defence mechanisms to keep hazardous substances out and to protect from their harmful effects. The respiratory system is protected by the sneeze reflex, nasal cavity, ciliary escalator and macrophages.
- Knowledge about routes of entry is used during the assessment of health risks and to determine appropriate control measures.
- Information about the substances can be gathered from product labels, material safety data sheets and exposure limit lists, although there are limitations with this information.
- Assessments sometimes require that basic surveys are carried out using equipment such as stain tube detectors, passive samplers, smoke tubes, dust monitoring equipment and dust lamps. There are some limitations in their use.

Routes of Entry

Hazardous substances enter the body through **absorption**. They can be absorbed through the skin, the lining of the lungs or the gastrointestinal tract.

The way a substance gets to these absorption locations is along a **route of entry**. Absorption may take place anywhere along the route.

Some substances can cause physical harm from contact, e.g. battery acid burning the skin from spillages. Others, such as epoxy resin, can sensitise from touching the skin.

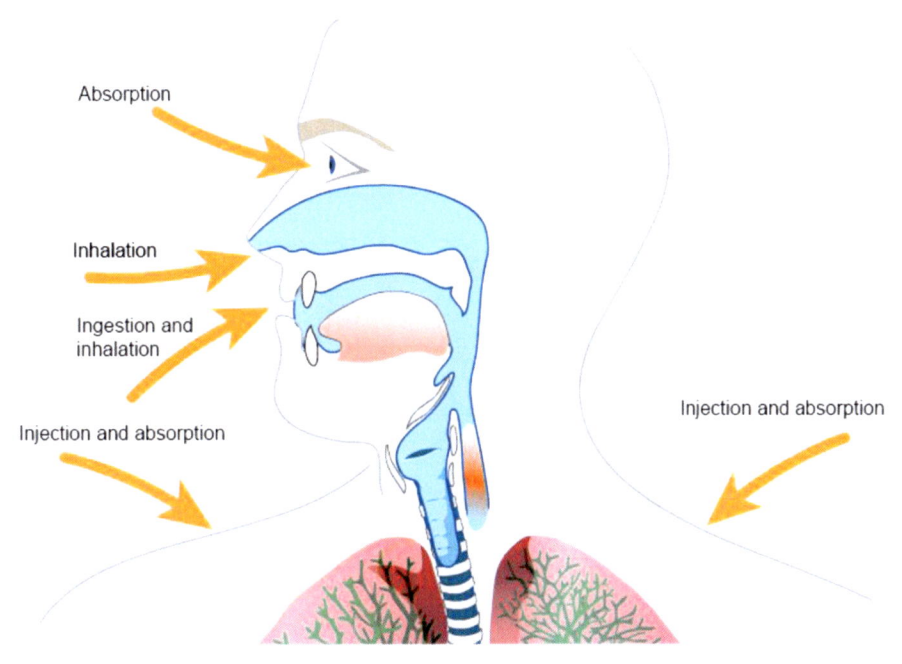

Routes of entry

12.2 Assessment of Health Risks

> **TOPIC FOCUS**
>
> There are four main **routes of entry** for hazardous substances into the body:
>
> - **Inhalation** - the substance is breathed in through the nose or mouth and travels along the respiratory passages to the lungs. The lung is the most vulnerable part of the body, as it can readily absorb gases, fumes, soluble dusts, mists and vapours. This is the main means of entry of biological agents.
>
>   There are two types of dust:
>
>   - Inhalable - particles of all sizes that can be inhaled into the nose and mouth and upper reaches of the respiratory tract.
>   - Respirable - particles smaller than 7 microns (0.007mm) that can travel deep into the lungs.
>
> - **Ingestion** - the substance is taken in through the mouth and swallowed, travelling the whole length of the gastrointestinal tract through the stomach and the intestines. This may occur:
>
>   - As a result of swallowing the agent directly.
>   - From eating or drinking contaminated foods.
>   - From eating with contaminated fingers.
>
>   All forms of chemicals may be ingested, and some biological agents may also enter the body by this route.
>
> - **Absorption** through the skin - the substance passes through the skin from direct contact with the agent or from contact with contaminated surfaces or clothing. It is mainly liquid chemicals which enter the body in this way, although other forms of chemical may either sufficiently damage the skin to gain entry or find their way through the eyes.
>
> - **Injection** through the skin - the substance enters directly into the body by high pressure equipment or contaminated sharp objects piercing the skin. Chemical liquids, and sometimes gases and vapours, may enter the body in this way. Biological agents are often injected - either on needles, etc. or by biting from an insect or infected animal.
>
> Although not a main route of entry, aspiration can also occur - where a substance already swallowed is regurgitated and can be inhaled into the lungs - usually if a person is unconscious.

Defence Mechanisms

The body's response against the invasion of substances likely to cause damage can be divided into **superficial** and **cellular** defence mechanisms.

Superficial

The skin provides a barrier against organisms and chemicals, but can only withstand limited physical damage. Some forms of dermatitis arise as a result of this damage, leading to thickening and inflammation of the skin which is both painful and unsightly.

The **respiratory tract** has a series of defences against inhaling contaminants:

- **The 'sneeze' reflex** - immediate irritation causing sneezing to expel contaminants.
- **Nasal cavity filters** - substances and micro-organisms down to 10 microns are trapped by nasal hairs and mucus.
- **Ciliary escalator** - the bronchioles, bronchi and trachea are lined with small hairs (cilia); mucus lining these passages is gradually brought up by these cilia out of the lungs. Particles above 7 microns trapped in the mucus are cleaned from the lungs by this mechanism.

Assessment of Health Risks | 12.2

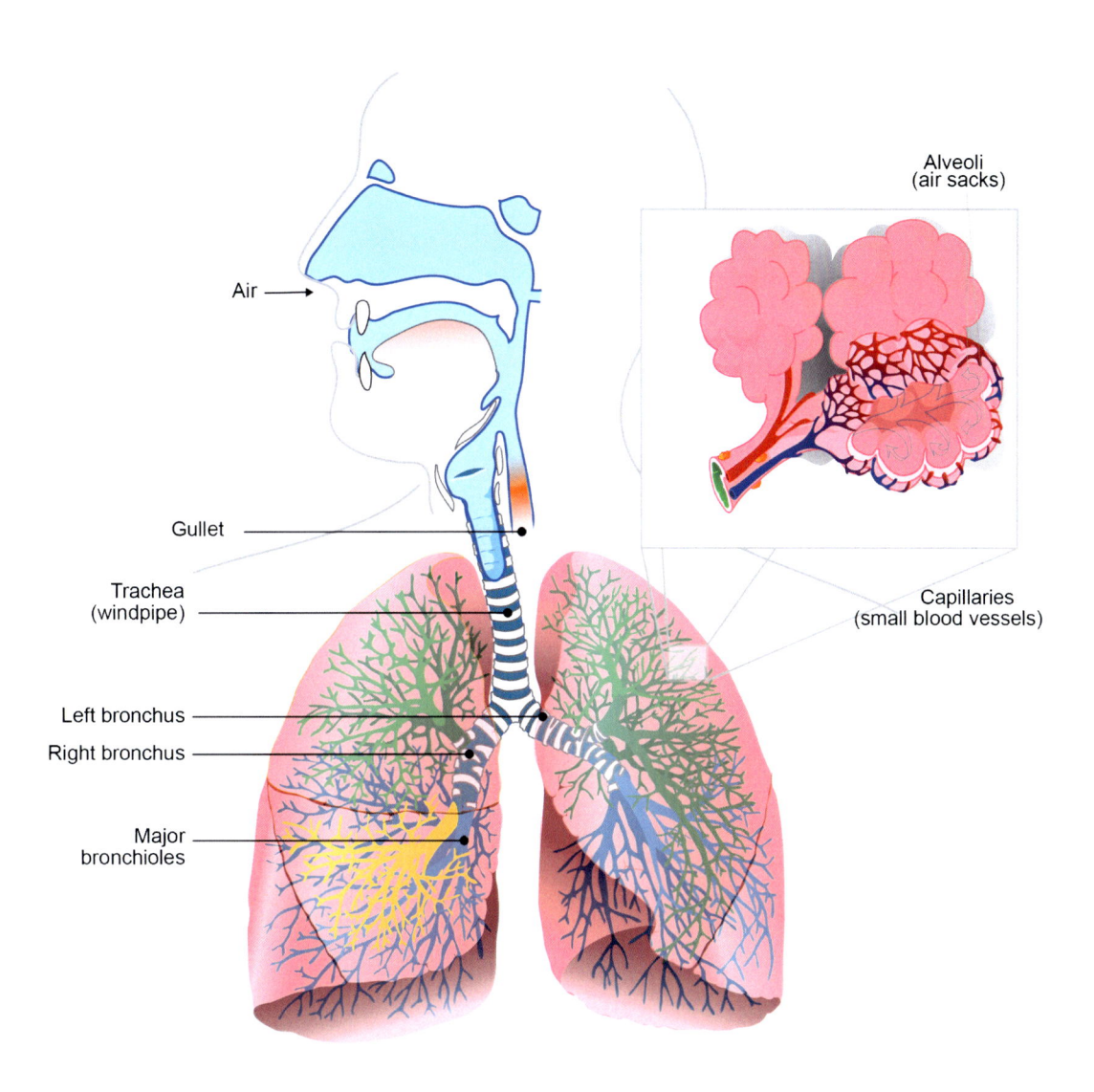

The respiratory system

Cellular

- **Macrophages** - scavenging white blood cells attack and destroy particles (fewer than seven microns) that lodge in the alveoli (the gas-exchange region in the lungs) where there are no cilia to protect them.
- **Inflammatory response** - any particles that cannot be removed by the macrophages are likely to trigger an inflammatory response, causing the walls of the alveoli to thicken and become fibrous. This can be temporary or result in permanent scarring (as with silicosis).
- **Prevention of excessive blood loss** - blood clotting and coagulation prevents excessive bleeding and slows or prevents the entry of contaminants into the blood.

12.2 Assessment of Health Risks

Factors that Need to be Taken into Account When Assessing Health Risks

Where there is a potential for construction workers to be exposed to hazardous substances, it will be necessary to assess that potential to ensure that harm does not occur. This is a requirement of the **Control of Substances Hazardous to Health Regulations 2002 (COSHH)**.

The risk assessment carried out to satisfy these regulations is often called a '**COSHH** Assessment'. There are five steps to **COSHH** assessment:

1. Gather information about the substance used, the people who might be exposed and the work activities carried out.
2. Evaluate the health risks - are current controls adequate?
3. Identify any further controls and implement them.
4. Record the risk assessment and actions taken.
5. Review and revise.

When identifying the hazardous substances on the construction site, remember that many are created by the work carried out, e.g. welding metal creates a metal fume; spraying paint creates an aerosol mist; these hazardous substances do not come pre-packaged and labelled, but are created by the construction work activities.

It will be seen later that information can be collected about hazardous substances by referring to various information sources. This information can be used to evaluate the health risks associated with the actual work practices.

TOPIC FOCUS

Factors to consider when carrying out an **assessment of health risks**:

- **Hazardous nature** of the substance - is it toxic, harmful, carcinogenic?
- **Physical form** of the substance - is it a solid, liquid, vapour or dust?
- The **quantity** of the hazardous substance present on site - including total amounts stored and the amounts actually in use or being created at any one time.
- Potential **ill-health effects** - will it cause minor ill health or very serious disease? And will this result from short-term or long-term exposure?
- **Duration** - how much exposure and for how long? Will it be for just a few minutes, or last all day?
- **Routes of entry** - will it be inhaled, swallowed, absorbed?
- **Concentration** - will a substance be used neat or diluted? What is the concentration in the air?
- **The number of people** potentially exposed and any vulnerable groups or individuals - such as expectant mothers or the infirm.
- **The control measures** that are already in place - such as ventilation systems and PPE.

MORE...

For more information on **COSHH**, visit:

www.hse.gov.uk/COSHH/index.htm

Assessment of Health Risks 12.2

All these factors have to be taken into account when doing the **COSHH** assessment, and then the adequacy of any existing control measures can be decided and additional controls and precautions selected.

Sources of Information

To assist the assessment of health risks, further information will be required. This can be obtained from product labels, safety data sheets and exposure limit documents.

Carcinogenic sign

> **DEFINITION**
>
> **SAFETY DATA SHEET**
>
> Provides all necessary information about the substance - for transport safety and to assist in carrying out the **COSHH** assessment.
>
> Note: Often wrongly called '**COSHH** Sheets' they are, in fact, nothing to do with the **COSHH Regulations** but rather relate to **REACH**.

The **Classification, Labelling and Packaging Regulation (CLP)** and the **European Registration, Evaluation, Authorisation and Restriction of Chemicals (REACH) Regulation** are the foundation of general chemicals legislation.

Product Labels

When supplying dangerous substances/mixtures, a product label must give the following information:

- Name, address and telephone number of the supplier.
- The nominal quantity of the substance/mixture (though this may be elsewhere on the package) - but only where it is made available to the general public.
- Product identifiers:
  - for substances: name and identification number (EC number, CAS number or inventory number);
  - for mixtures: trade name, and the identity of all the substances (maximum of 4) in the mixture which contribute to its classification.
- Hazard pictograms.
- Signal word (as applicable).
- Hazard statements (as applicable).
- Precautionary statements (as applicable).
- Supplementary information.

12.2 Assessment of Health Risks

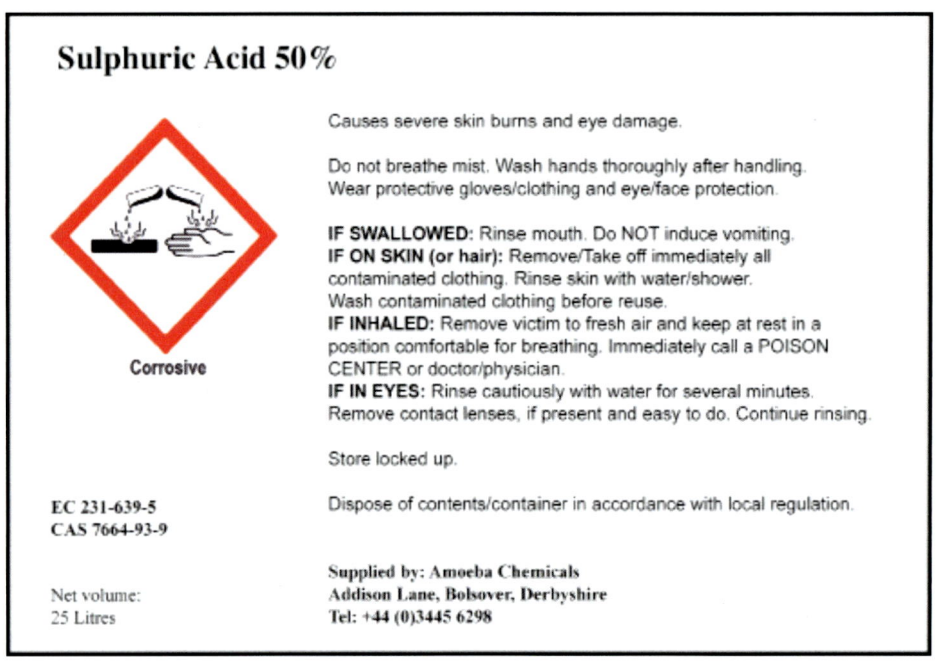

A label showing the key information about the hazardous nature of the product

Safety Data Sheets

Article 31 of **REACH** requires suppliers of dangerous substances and preparations to provide safety data sheets.

Safety data sheets are intended to provide users with sufficient information about the hazards of a substance or preparation for them to take appropriate steps to ensure health and safety in the workplace in relation to all aspects of its use, including its handling, transport and disposal.

They are not COSHH Assessments and should not be taken as such.

Assessment of Health Risks | 12.2

TOPIC FOCUS

Safety data sheets contain the following information:

- Identification of the substance or preparation, and supplier - name, address and emergency contact phone numbers.
- Hazard identification - a summary of the most important features, including likely adverse human health effects and symptoms.
- Composition and information on ingredients - chemical names, classification code letters and risk phrases.
- First-aid measures - separated for the various risks, and specific, practical and easily understood.
- Fire-fighting measures - emphasising any special requirements.
- Accidental release measures - covering safety, environmental protection and clean-up.
- Handling and storage - recommendations for best practice, including any special storage conditions or incompatible materials.
- Exposure controls and personal protection - any specific recommendations, such as particular ventilation systems and PPE.
- Physical and chemical properties - physical, stability and solubility properties.
- Stability and reactivity - conditions and materials to avoid.
- Toxicological information - acute and chronic effects, routes of entry and symptoms.
- Ecological information - environmental effects of the chemical, which could include patterns of degradation and effects on aquatic, soil and terrestrial organisms, etc.
- Disposal considerations - advice on specific dangers and legislation.
- Transport information - special precautions.
- Regulatory information - e.g. labelling and any relevant national laws.
- Other information - e.g. list of relevant risk phrases, any restrictions on use (non-statutory supplier recommendations).

Safety data sheets must be supplied (paper or electronic) free of charge when the substance is first provided. They must be kept up to date and revised and reissued accordingly.

Limitations of Information Used When Assessing Risks to Health

The sources of information we have seen are important, but have limitations in assessing health risks:

- They contain general statements of the hazards, but do not take into account local conditions in which you will use the substances, which will affect the risk.
- The information can be very technical and difficult to understand by the non-specialist.
- Substances affect different people in different ways - this is not taken into account in the generalities used.
- Information is about a substance or preparation in isolation - no account is taken of the effects of mixed exposures.
- The information was good at the time it was written; it represents current scientific thinking, so there may be hazards present that are not currently understood.

Role and Limitations of Hazardous Substance Monitoring

Hazardous substance monitoring sets out to measure how much of a contaminant is in the air (inhalation is the only route of entry that we can positively measure), and we use this, together with time exposure, to assess the risks to health of substance exposure.

To carry out hazardous substance monitoring, we use various types of sampling equipment to collect and measure how much contaminant is in the air.

> **TOPIC FOCUS**
>
> ### Sampling Techniques
>
> The first task in our basic survey is to collect the sample of air so that it may be analysed. We need to consider (depending on the risk level of the contaminant being assessed):
>
> - **Location of the sample** - it may be taken in the general working atmosphere, in the operator's breathing zone, or at a position close to the contaminant generation or use.
> - **Method of analysis** - this may involve sampling and analysis in the same instrument, or taking the sample collected and analysing it using different equipment, perhaps in a laboratory away from the point of collection.
> - **Duration of the sampling** - is the survey looking at short- or long-term exposures?

Hazardous substance monitoring surveys generally fall into three main categories:

- A **spot** or **grab** sample - a snapshot of airborne concentration at one moment in time - usually analysed on the spot.
- A better method of obtaining a time-weighted average is by **collecting a sample over a period** and then analysing it. This is the usual technique for personal monitoring.
- A **continuous monitored** sample (usually high-risk areas) - where a sample is collected and continuously analysed over a period of time. Such systems may be linked to an alarm system if safe levels are exceeded.

There are two basic **methods of sampling**, based on the way in which the sample is collected:

- **Diffusion** or **passive sampling** - where the air sample (along with any contaminants in it) passes over the sampling system naturally, through an absorbent material which can be removed for later analysis.
- **Mechanical** or **active sampling** - where a pump forces air flow through the sampling device - used for both spot and continuous sampling.

Stain Tube Detectors

These are easy to use and useful for analysing gas and vapour contamination in air at one moment in time (spot sampling).

The principle of operation is simple - a known volume of air is drawn over a chemical reagent contained in a glass tube. The contaminant reacts with the reagent and a coloured stain is produced. The degree of staining can give a direct reading of concentration.

The instrument comprises a glass tube containing the chemical reagent fitted to a hand-operated bellows pump or piston-type pump. Many types of tube are available, with different chemicals that react to different gases and vapours.

To operate:

- Select the appropriate tube.
- Snap off the end of the tube to open it.
- Place the open end on the pump and break off the other end.
- Squeeze the bellows or operate the pump for a specified amount (e.g. number of squeezes of the bellows).

This draws air through the detector tube, the chemical in the tube changes colour and the concentration of the contaminant can be read from a scale marked along the tube.

The following diagram illustrates the principle:

Stain tube before and after use. Note the closed and open ends of the tube. Arrow shows direction of air flow. n = 10 indicates that 10 strokes on the hand bellows are required. These tubes are sensitive to carbon monoxide (CO). Final concentration is given as 50 parts-per-million (ppm)

Limitations of stain tube detectors:

- Provide a spot-sample for one moment in time rather than an average reading.
- Can have an accuracy of +/-25%, which is not particularly accurate.
- Correct number of strokes must be used; losing count and giving too few/too many will give inaccurate results.
- Volume of air sampled may not be accurate due to incorrect assembly interfering with the air flow (through leaks, etc.) or incorrect operation.
- Can be cross-sensitive to substances other than the one being tested for.
- Designed to operate at about 20°C and one atmosphere pressure. Problems may be caused by variations in temperature and pressure away from these standard conditions.
- Tubes have a shelf storage life; out-of-date tubes may be inaccurate due to deterioration of the reagent.
- There may be variations in the precise reagent make-up between tubes.

12.2 Assessment of Health Risks

Passive Samplers

These use absorbent material to sample contaminants without using a pump to draw air through the collector. They give a measure of concentration over a period of time (long-term sampling) and can be used for gas or vapour. There are two main types of design:

- The **badge (or dish) sampler** has a flat, permeable membrane supported over a shallow layer of sorbent.
- The **tube-type sampler** has a smaller permeable membrane supported over a deep metal tube filled with sorbent.

They allow gas or vapour to diffuse to an absorbent surface. At the end of sampling, the sampler is sent for laboratory analysis, although some work on a colour-change principle similar to litmus paper. Working on a colour-change principle allows visual assessment against a standard chart.

Limitations of passive samplers:

- Do not provide any immediate indication of the contamination concentration - results have to be analysed (unless colour changing).
- Only measure accumulated concentrations over the period for which they are in use - cannot be easily used to calculate time-weighted averages.
- Only sample contamination where they are located or, in the case of badges, where the wearer is - cannot be easily used to take spot-samples in various parts of the workplace.
- Badges are easy to take off, rendering them ineffective.
- Size of the sample is imperative. If the samplers are only used intermittently or only a small sample is used, results may be misleading.

Badge sampler (Retaining ring, Permeable membrane, Sorbent)

Tube sampler (Permeable membrane, Sorbent, Metal tube)

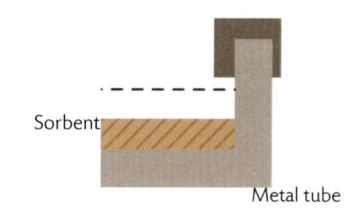

Oxygen Meters

Direct reading instruments are available to monitor and warn of concentrations of the toxic gases carbon dioxide, carbon monoxide or hydrogen sulphide in the atmosphere, as well as the essential gas, oxygen, in naturally occurring respirable air. They are usually reliable and accurate.

They indicate concentrations of oxygen in the atmosphere on a simple dial or digital readout. To do this, an air sample diffuses into the sensor through a special membrane, where the resultant electrochemical process produces electric current directly proportional to the oxygen concentration.

These instruments can be pre-set to a given oxygen concentration which activates an audible or visual alarm system. They are used both as personal monitors and to measure room concentrations of oxygen (e.g. in a confined space).

The normal percentage of oxygen in air is 21%, most of the remainder (78%) being nitrogen.

Limitations of oxygen meters:

- Sensitive (but accurate) sensors, so are sometimes delicate and need careful handling.
- Need some skill to accurately set and monitor.
- Battery needs to be proven for capacity before use.

Smoke Tubes

These are simple devices that generate non-toxic smoke in a controlled chemical reaction. They are similar in appearance and operation to stain tubes, and to operate you break open the tube and attach a rubber bulb to emit the smoke.

Smoke tubes are used to assess the strength and direction of airflow. The smoke they release is carried away by the air currents in the local environment and the movements observed. Such smoke tests are ideal for checking the effectiveness of ventilation and extraction systems, air-conditioning systems and chimneys. They can be used to detect leaks in industrial equipment, to assess relative air pressures used in certain types of local ventilation systems and to provide general information about air movements in the general work area.

Limitations of smoke tubes:

- Do not give a quantitative measure of concentration, only a qualitative indication of air movement.
- Smoke particle size will probably be different from the size of contaminant particles, so the smoke may move in air currents in a different way.

Dust Monitoring Equipment

Dust Lamp (Tyndall Lamp)

Airborne dust in the workplace which is not visible to the naked eye can be visualised using a dust lamp.

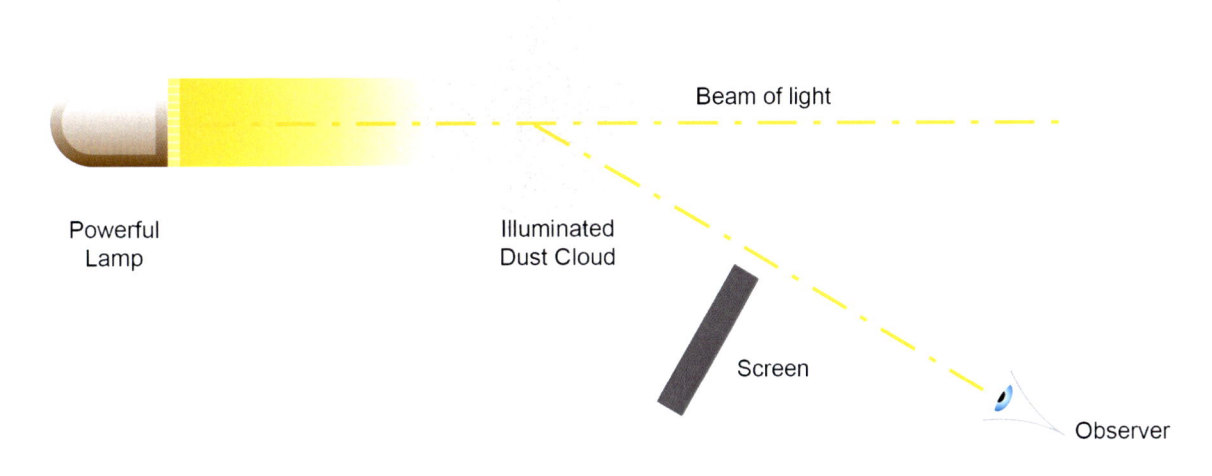

Tyndall beam apparatus

A strong beam of light is shone through the area where a cloud of finely divided dust is suspected. The eye of the observer is shielded from the light beam and the dust cloud is made visible. The method is used to determine how exhaust ventilation systems are working.

Limitations of dust lamps:

- Do not provide numerical data, only a qualitative indication.
- Provide no differentiation between contaminants and any other dusts.

12.2 Assessment of Health Risks

Indirect-Reading Filtration Equipment (Dust Sampler)

Dust exposure can be measured using a sampling train made up of an air pump, tube and sampling head. This can be worn by a worker to give a personal sample covering their work period, or placed in a static location to get a background sample.

Air is drawn through and dust is collected onto a pre-weighed filter; it is sent to a laboratory where it is weighed again and the amount of dust in the air calculated. This will be an average value over the chosen period of time.

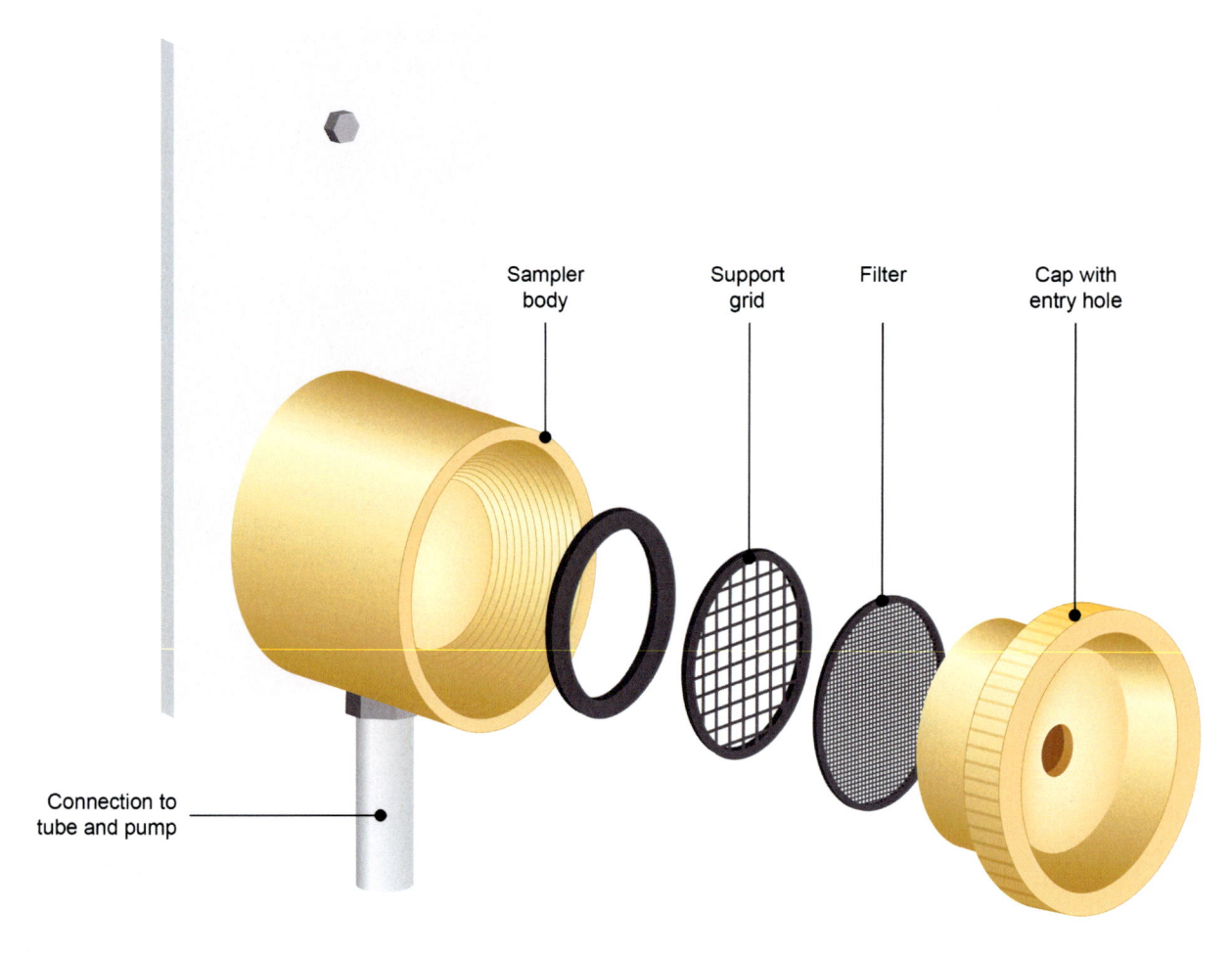

Dust sampler head showing filter in position
Based on original source MDHS 14/4 General methods for sampling and gravimetric analysis of respirable, thoracic and inhalable aerosols, HSE, 2014 (www.hse.gov.uk/pubns/mdhs/pdfs/mdhs14-4.pdf)

Limitations of dust samplers:

- Only suitable for calculating average exposures over long periods (minimum four hours).
- All the dust is assumed to be the contaminant dust for calculation purposes (unless more extensive analysis is carried out).
- Easy to misuse.

Purpose of Occupational Exposure Limits and How They Are Used

Occupational Exposure Limits (OELs) help to control exposure to hazardous substances in the workplace by defining the maximum amount of (air) concentration of a substance that can safely be allowed.

Limit values are laid down throughout the EU, but each Member State is able to establish its own national OELs, often going beyond EU legislation.

All OELs assume that the workers who are exposed are healthy adults, although in some cases they also aim to protect 'sensitive groups'. Normally, exposure limits do not apply to pregnant women and nursing mothers, for example, and specific action should be taken where necessary to protect these groups.

In the UK, legally enforceable OELs exist under **COSHH** as WELs and are published in the Health and Safety Executive (HSE) document EH40 *Workplace exposure limits*.

DEFINITION

WORKPLACE EXPOSURE LIMITS (WELs)

Maximum concentrations of airborne contaminants, averaged across a particular reference period of time, to which employees may be exposed.

The purpose of WELs is to put a ceiling in place so that employees will not be exposed to concentrations of airborne substances (either for short durations of time or for long periods of the working day) where scientific evidence suggests that there is a risk to their health.

WELs have legal status under **COSHH** and can be found listed in the HSE Guidance Note EH40. If a WEL is exceeded, then a breach of the **COSHH** has taken place; this might lead to enforcement action or prosecution.

TOPIC FOCUS

Units of Measurement of Exposure

The two main units used for measuring airborne concentrations are:

- Parts per million (ppm).
- Milligrams per cubic metre of air (mg/m^3, or $mg\ m^{-3}$).

Vapours and gases are measured in ppm (parts per million), which refers to the number of parts of vapour or gas of a substance in a million parts of air by volume. Particulate matter - dusts, fumes, etc. - is measured in mg/m^3, which refers to the milligrams of the substance per cubic metre of air.

One further unit of measurement is used in relation to fibres (e.g. asbestos):

- Concentrations of fibres are expressed in fibres per millilitre of air (fibres ml^{-1}).

Long-Term and Short-Term Limits

The two reference periods commonly used for workplace exposure limits are:

- **15 minutes** - short-term exposure limit.
- **8 hours** - long-term exposure limit.

12.2 Assessment of Health Risks

The reasons for having two limits are:

- **Short-Term Exposure Limits (SLTELs)** combat the ill-health effects (acute effects) of being exposed to very high levels of the substance for quite short periods of time.
- **Long-Term Exposure Limits (LTELs)** combat the ill-health effects of being exposed to relatively low concentrations of a substance for many or all hours of every working day throughout a working lifetime (chronic effects).

Significance of Time-Weighted Averages

Workplace exposure limits are Time-Weighted Average (TWA) exposures. They are calculated by measuring a person's average exposure over a specific reference period of time, either **15 minutes** (short-term exposure limit) or **8 hours** (long-term exposure limit). Consequently, they do not provide a limit for airborne concentrations measured over a very short period of time (say 1 or 2 seconds). Although such instantaneous measurements are useful as part of a monitoring programme to identify peak concentrations, only time-weighted averages can be used to legally assess exposure against WELs.

Limitations of Exposure Limits

WELs are designed to control the absorption into the body of harmful substances after they have been inhaled. They are not concerned with absorption from swallowing or through contact with the skin or eyes. So they will not take into account high levels of solvent that are present in a person who has had skin contact with it for a period of time - only that which they have inhaled.

They take no account of human sensitivity or susceptibility. This is particularly important in the case of substances which produce an allergic response - once a person has become sensitised, the exposure limit designed to suit the average person has no further validity.

Other limitations include:

- They do not take account of the synergistic (combined) effects of mixtures of substances, e.g. the use of multiple substances.
- They do not provide a clear distinction between 'safe' and 'dangerous' conditions.
- They cannot be applied directly to working periods which exceed 8 hours.
- They may become invalid if the normal environmental conditions are changed, e.g. changes in temperature, humidity or pressure may increase the harmful potential of a substance.

Application of Workplace Exposure Limits

WELs are designed to control absorption into the body of airborne harmful substances following inhalation. EH40 contains the list of substances for which WELs have been set, together with the LTEL and STEL values of these WELs, and can be used to determine whether exposure to an airborne contaminant has been adequately controlled as required by the **COSHH**. Adequate control of exposure to an airborne contaminant which is hazardous to health means not exceeding the WEL, or for a substance that is carcinogenic, mutagenic or causes asthma, reducing exposure to as low as is reasonably practicable.

Calculating WELs: A Worked Example

Workers have complained about breathing difficulties in an area where paint is being applied. The workers have said that there is a strong, sweet (not at all unpleasant) odour.

A paint is being applied that contains ethyl acetate. The WEL is 200ppm, 8-hour TWA and 400ppm, 15 minutes STEL. These values are obtained from safety data sheets and/or EH40.

Air samples have been taken in the area where the paint is being applied and the following results have been recorded.

Assessment of Health Risks 12.2

Time	ppm	Total Exposure ppm	Exposure TWA
09:00 to 10.00	100	100	
10:00 to 11:30	200	300	
11:30 to 13:00	200	300	
13:00 to 14:00	100	100	
14:00 to 16:00	250	500	
16:00 to 17:00	100	100	
		1400 ppm	1400/8=175ppm

The TWA from the sample survey shows that 175ppm per hour average. This means that the 8-hour TWA WEL has not been exceeded.

The STEL absolute limit for 15 minutes is 400. The time period 14:00 to 16:00 has recorded 500ppm. The person undertaking the survey would need to establish if this were a single sample and the circumstances behind such a high reading. The process must not continue under these circumstances.

Principle of Reducing Exposure Levels

The operation of WELs is based on controlling risk by reducing the workplace exposure to the contaminant. Therefore, the **COSHH** require that exposure to harmful substances should be reduced to the lowest level reasonably practicable:

- Eliminating exposure is the best way to control risk. Although this has been adopted for certain chemicals (e.g. carcinogens), it is impractical in most situations when we take into account the requirements of working processes. Therefore, limitation of the risk becomes the next best strategy.

- In practice, reducing exposure may mean more than simple compliance with the stated WEL. Under the **COSHH** requirement, if it is reasonably practicable to get contamination levels even lower, then that standard should be achieved.

STUDY QUESTIONS

5. Identify the routes of entry of chemicals into the body.
6. What is the difference between an inhalable substance and a respirable substance?
7. What information is generally provided on the product label of a substance that is classified as dangerous for supply?
8. What is the purpose of safety data sheets?
9. What is the difference between passive and active sampling devices?
10. Give three limitations in the use of stain tube detectors.
11. What are smoke tubes used for?
12. What is Guidance Note EH40?
13. What is a Workplace Exposure Limit (WEL)?
14. What do you understand by the term 'time-weighted average' in relation to a WEL?
15. Give three examples of the limitations of WELs.
16. What two reference periods are commonly used with TWAs?

(Suggested Answers are at the end.)

12.3 Control Measures

Control Measures

IN THIS SECTION...

- Exposure to hazardous substances should be prevented or, if that's not possible, controlled to below the workplace exposure limits.
- There are principles of good practice for the control of exposure: minimisation of emissions; consider routes of exposure; appropriate and effective controls; use of PPE; checks on effectiveness of controls; provision of information and training; and ensuring controls do not increase risk to health and safety.
- Measures to achieve the principles of good practice include: eliminate or substitute the substance; change the process; reduce exposure time; enclose or segregate; local exhaust ventilation; dilution ventilation; respiratory protective equipment; other PPE; personal hygiene; and health surveillance.
- Further controls may be required for substances that can cause cancer, asthma or damage to genes that can be passed from one generation to another.

The Need to Prevent Exposure

COSHH require the employer to prevent exposure to substances hazardous to health if it is reasonably practicable to do so. The employer might:

- Change the process or activity so that the hazardous substance is not needed or generated.
- Replace the substance with a safer alternative.
- Use the substance in a safer form, e.g. pellets instead of powder.

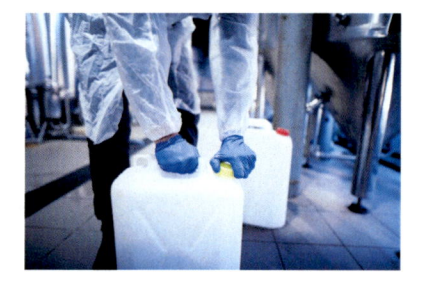

Adequately Control Exposure

If prevention is not reasonably practicable, exposure is to be adequately controlled. This will require putting in place measures appropriate to the activity and consistent with the risk assessment, following the hierarchy of controls.

Under the **ISO 45001** and **COSHH** the general hierarchy of control is to:

- Eliminate or substitute the hazard by using a less hazardous agent.
- Change the process, i.e. vacuum instead of brush.
- Reduce the time of exposure by providing regular breaks.
- Use physical or engineering controls to reduce the risk at source and provide general protection (segregation, enclosure, ventilation).
- Manage the task or person by job design and provide (as a last resort) personal protective equipment.

> **DEFINITION**
>
> **ADEQUATE CONTROL**
>
> Under **COSHH**, adequate control of exposure to a substance hazardous to health means:
>
> - applying the eight principles of good practice set out in Schedule 2A of **COSHH**;
> - not exceeding the WEL for the substance (if there is one); and
> - if the substance causes cancer, heritable genetic damage or asthma, reducing exposure to as low as is reasonably practicable.

Ensuring WELs Are Not Exceeded

The HSE has established WELs for a number of substances hazardous to health. These are intended to prevent excessive exposure to the substance by containing exposure to below a set limit. Correctly applying the **principles of good practice** will mean exposures are controlled below the WEL.

Principles of Good Practice

There are eight principles of good practice, detailed in **COSHH**. These are outlined below:

- Principle 1

    Design and operate processes and activities to minimise emission, release and spread of substances hazardous to health.

- Principle 2

    Take into account all relevant routes of exposure - inhalation, skin and ingestion - when developing control measures.

- Principle 3

    Control exposure by measures that are proportional to the health risk.

- Principle 4

    Choose the most effective and reliable control options that minimise the escape and spread of substances hazardous to health.

- Principle 5

    Where adequate control of exposure cannot be achieved by other means, provide, in combination with other control measures, suitable PPE.

- Principle 6

    Check and review regularly all elements of control measures for their continuing effectiveness.

- Principle 7

    Inform and train all employees on the hazards and risks from substances with which they work, and the use of control measures developed to minimise the risks.

- Principle 8

    Ensure that the introduction of measures to control exposure does not increase the overall risk to health and safety.

Common Measures Used to Implement the Principles of Good Practice

Elimination or Substitution of Hazardous Substances

It may be possible to eliminate or substitute the substance by:

- Eliminating the process or type of work that requires the use of (or creates) the substance (e.g. outsourcing a paint-spraying operation).
- Changing the way that the work is done to avoid the need for the substance (e.g. screwing items together rather than gluing).
- Disposing of unused stock of substances that are no longer needed.
- Substituting the hazardous substance for one non-hazardous (e.g. switch from an irritant to a non-hazardous floor cleaner).

12.3 Control Measures

- Substituting a hazardous substance for one that has a lower-hazard classification (e.g. exchange a solvent paint for a water-based paint).
- Changing the physical form of a substance (e.g. use pellets instead of powder).

Process Changes

It may be possible to change a process so that risks can be reduced, for example:

- Brush painting rather than spraying reduces airborne mist and vapour.
- Vacuuming, rather than sweeping up, reduces dust levels.
- Damping of a substance during mixing or when clearing up also reduces dust levels.

Reduced Time Exposure

Ill-health effects caused by hazardous substances are often related to the length of time of exposure and the dose (amount) of the contaminant. Reducing the time will reduce the dose (extending the time increases the dose). We should therefore look to minimise the time people work with hazardous substances, especially with those having acute effects. If a short-term exposure limit (15-minute TWA) exists for the substance, this must not be exceeded. We can achieve this by:

- Providing regular breaks away from contact with the hazardous substance.
- Job rotation - where exposure of an individual is reduced by sharing the dose with other workers.

Enclosure and Segregation

Where it is not possible to reduce exposure, then we have to consider physical controls which enclose the hazard and segregate people from the process involving it.

Total enclosure of a process which generates dust or fumes will prevent the escape of airborne contaminants which could be inhaled by operators nearby. However, it may still be necessary to access equipment or material within that area, so the use of robotically controlled, remote handling systems may be incorporated, allowing access without disturbing the integrity of the enclosure.

Where isolation of the source is difficult, it may be more practical to enclose the workers to ensure that they remain segregated from the hazard (e.g. in a control booth).

Local Exhaust Ventilation

A Local Exhaust Ventilation (LEV) system will contain and collect dusts, vapours and fumes where they are generated, and prevent them spreading further into the workplace. The contaminants will be filtered out and the clean air exhausted outside the workplace.

Control Measures | 12.3

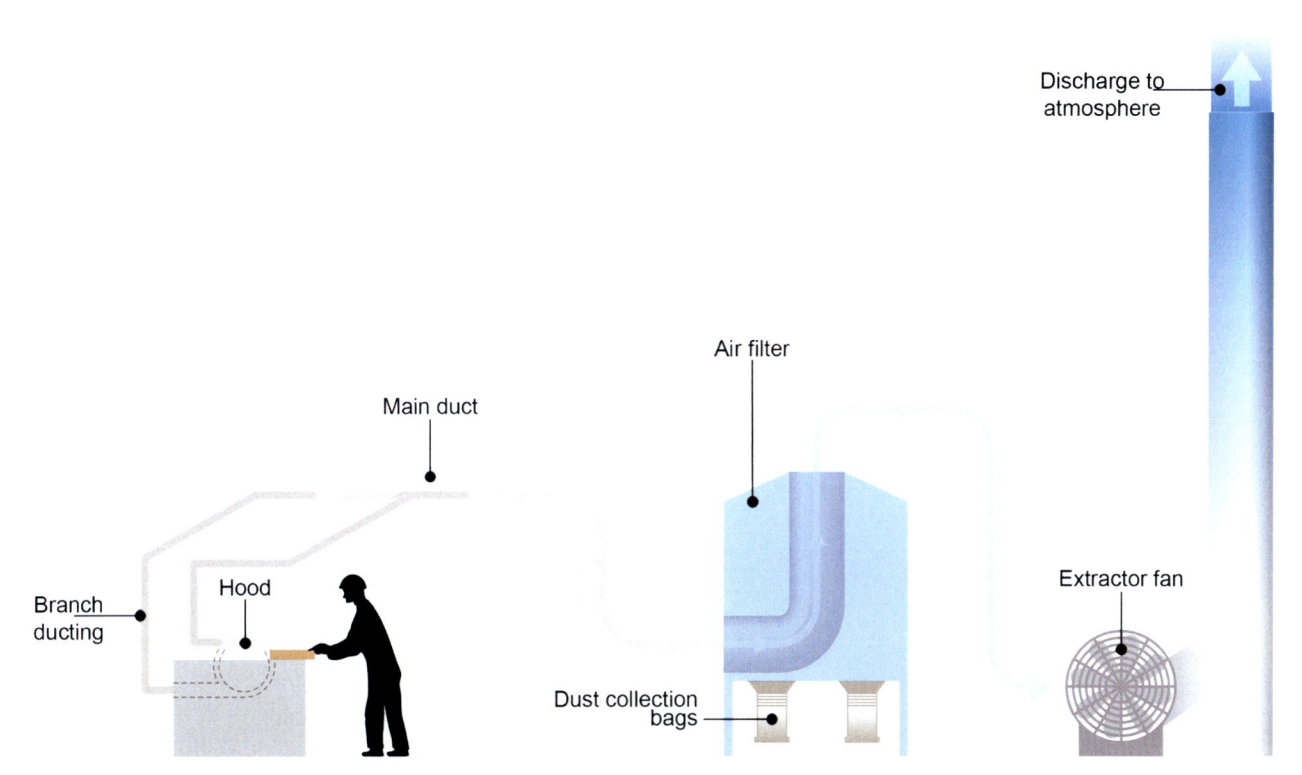

A typical LEV system extracting sawdust from a bench-mounted circular saw

TOPIC FOCUS

A **typical LEV system** consists of:

- An **intake hood** that draws air containing the contaminant in at the point it is created.
- **Ductwork** that carries the air from the intake hood.
- A **filter system** that cleans the contaminant from the air to an acceptable level.
- A **fan** that provides the air movement through the system.
- An **exhaust duct** that discharges the clean air to atmosphere.

Examples of LEV include:

- **Glove boxes** - total enclosures, often used in laboratories, which are accessed through flexible gloves and kept under negative pressure to prevent any release of contaminant.
- **Fume cupboards** - partial enclosures, again used in laboratories, accessed through a vertical sliding sash, with the enclosure kept under negative pressure so that the airflow is through the sash into the hood to prevent any release of contaminant.

A variety of different intake hoods are used on LEV systems, but they can be categorised into two main types:

- **Captor hoods** - hoods which can be positioned as near as possible to the hazard and capture contaminants by a negative airflow into the hood before they reach the operator, e.g. those used to extract woodworking dust or welding fume.
- **Receptor hoods** - large structures designed to capture contaminants which are being directed naturally into the hood, so that less air movement is needed to achieve uptake (e.g. a large intake hood above a bath of molten metal - the metal fume will be hot and rising up into the hood on convection currents).

12.3 Control Measures

MORE...

For more information on LEVs, visit:

www.hse.gov.uk/lev/index.htm

Factors that Reduce Effectiveness

The effectiveness of LEV systems can be affected by:

- Poorly positioned intake hoods.
- Damaged or leaking ducts.
- Excessive amounts of contamination.
- Ineffective fan due to slow speed or lack of maintenance.
- Blocked filters.
- Build-up of contaminants in the duct.
- Sharp bends in the duct.
- Unauthorised additions into the system.

Inspection and Monitoring

LEV systems must be routinely inspected and maintained to ensure their continuing effectiveness:

- **Regular Visual Inspections**
    - To check the integrity of the system, signs of obvious damage and build-up of contaminant inside and outside the ductwork.
    - Filters should be regularly checked to ensure they are not blocked (some have a collector can which can be emptied).
    - The exhaust outlet should be clear.
- **Planned Preventive Maintenance**

    May include:
    - Replacing filters.
    - Lubricating fan bearings.
    - Inspecting the fan motor.
- **Periodic Testing**

    To ensure that air velocities through the system remain adequate:
    - Can be done by visual inspection of the captive system using a smoke-stick, measuring air velocity at the intake and along the ductwork using an anemometer, and measuring static pressures with manometers and pressure gauges.

LEV provided as a control measure for **COSHH Regulations** substances should be thoroughly examined by a competent person every 14 months or otherwise in accordance with Schedule 4 of **COSHH Regulations**: "*Frequency of thorough examination and test of local exhaust ventilation plant used in certain processes*".

MORE...

Further information on LEV can be found in HSG258 *Controlling airborne contaminants at work: A guide to local exhaust ventilation (LEV)*, available at:

www.hse.gov.uk/pubns/books/hsg258.htm

Dilution Ventilation

This operates by diluting the contaminant concentration in the general atmosphere to an acceptable level, by efficiently changing the air in the workplace over a given period of time, e.g. a number of complete changes every hour.

The air changes might be achieved:

- **Passively** - by providing low-level and/or high-level opening louvres.
- **Actively** - using powered fans.

This type of ventilation is intended to be effective in removing gas contaminants (sometimes fumes) and to keep overall concentration of any contaminants below the WEL.

Dilution ventilation is appropriate where:

- The WEL of the hazardous substance is high.
- The rate of formation of the gas or vapour is slow.
- Operators are not in close contact with the contamination generation point.

If a powered system is used, fans must be sited appropriately. If the contaminant is:

- **Lighter** than air, it will naturally rise up inside workrooms and be extracted at high level.
- **Heavier** than air, it will sink to the floor and low-level extraction will be more suitable.

Limitations of dilution ventilation systems:

- Not suitable for the control of substances with high toxicity.
- Do not cope well with sudden releases of large quantities of contaminant.
- Do not work well:
  - On dust.
  - Where the contaminant is released from one point source.
- Dead areas may exist where high concentrations of the contaminant are allowed to accumulate.

    Dead areas are those areas in a workplace which remain dormant, so the air in them is not changed. This is usually due to the air-flow patterns produced by poor positioning of extraction fans and inlets for make-up air. Dead areas can move from one place to another as a result of changing the positions of fans and make-up air inlets, or by the intrusion of other air through windows and doors. Moving the position of machinery or workbenches can have similar effects.

Respiratory Protective Equipment

Purpose and Application

> **DEFINITION**
>
> **RESPIRATORY PROTECTIVE EQUIPMENT (RPE)**
>
> Any type of PPE specifically designed to protect the respiratory system, e.g. self-contained breathing apparatus.

The general principles of PPE can be applied to Respiratory Protective Equipment (RPE), in that it is worn by workers to reduce the possibility of harm from exposure to a hazardous substance. This is called a **safe person strategy**. Ideally, the safe person strategy is a second line of defence against a potential hazard - control at source, or a **safe place strategy**, should be the first aim.

12.3 Control Measures

There will be situations where personal protection is the most appropriate method to deal with a particular hazard, e.g. when the cost of controlling the hazard at source is high and the time required for protection is short.

Classic situations which typify these conditions are:

- Work involving planned maintenance, e.g. during plant shut-downs.
- One-off tasks generating airborne contaminants, e.g. demolition of a building by pulling it down.
- Failure of primary safety systems or emergency situations, e.g. a chemical leak from an on-site storage tank.

Types of RPE and their Effectiveness

There are two main categories of RPE:

- **Respirators** - designed to filter the air from the immediate environment around the wearer.
- **Breathing apparatus** - provides breathable air from a separate source.

We will now go on to describe the different types of equipment and their effectiveness.

Respirators

These come in a variety of types:

- **Filtering Facepiece Respirator**

    This is the simplest type, consisting of a piece of filtering material worn over the nose and mouth and secured by elastic headbands. Fit around the chin and face depends on the tension in the headbands; a flexible metal strip enables the user to bend it over the bridge of the nose to ensure a personal fit.

    This type of respirator is useful to prevent inhalation of dust or fibres (and sometimes gas and vapours), but is not suitable for high concentrations of contaminant, for use against substances with high toxicity, or for long-duration use.

A worker using a filtering facepiece respirator to prevent inhalation of wood dust

Use and Benefits	Limitations
Cheap.	Low level of protection.
Easy to use.	Does not seal against the face effectively.
Disposable.	Uncomfortable to wear.

- **Half-Mask or Ori-Nasal Respirator**

    A flexible rubber or plastic facepiece which covers the nose and mouth, with one or two filtering canisters (cartridges) that contain the filtering material. It gives a much higher level of protection than the filtering facepiece respirator.

    When the wearer inhales, a negative pressure is created inside the facepiece; this means that any leak in the respirator or around the seal will allow contaminants in.

A worker wears a half-mask respirator to seal asbestos lagging around a pipe

Control Measures | 12.3

Use and Benefits	Limitations
Good level of filtration.	No built-in eye protection.
Good fit achievable.	Negative pressure inside facepiece.
Easy to use.	Uncomfortable to wear.

- **Full-Face Respirator**

    This is similar to the half-mask (also with canister filters) but has a built-in visor that seals in the eyes and face. This type gives a high level of protection against airborne contaminants and protects the eyes. This can be important where contaminants may splash, cause irritation or be absorbed through the eyes.

Use and Benefits	Limitations
Good level of filtration.	Restricts vision.
Good fit achievable.	Negative pressure inside facepiece.
Protects the eyes.	Uncomfortable to wear.

A full-face respirator with filtering canister (or cartridge)

- **Powered Visor Respirator**
    - A powered fan blows filtered air to the wearer.
    - Usually made up of a helmet and face visor, with the air drawn in through a filter in the helmet, and fed down over the face inside the facepiece.
    - Powered by rechargeable battery.

    This type of respirator does not have a tight seal around the face, and is especially suited to dusty, hot environments where the stream of air moving over the face is a benefit. Similar is the **powered clean air respirator** which has the filter remote from the visor, usually worn on the belt, and fed to the visor through a tube.

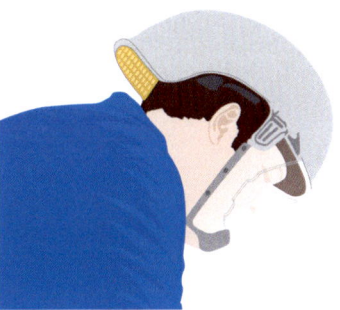

Use and Benefits	Limitations
Intermediate level of filtration.	Heavy to wear.
Air movement cools wearer.	No tight face seal.
Air stream prevents inward leaks.	Limited battery life.

Powered visor respirator

Breathing Apparatus (BA)

This can be classified under three general headings:

- **Fresh-Air Hose BA**

    This is the simplest type, where a large diameter hose is connected to the user's face mask. Air is either drawn down the hose by breathing or blown down by a low speed/low pressure fan.

Use and Benefits	Limitations
Air is from outside the workroom.	Hose must be tethered.
Supply of air is not time-restricted.	Bends or kinks make breathing difficult.
	User is restricted by length of hose.

12.3 Control Measures

- **Compressed Air BA**

Similar to the fresh-air hose BA, but air is supplied through a small-bore hose at high pressure. Pressure is stepped-down by a regulator and supplied at low pressure to the user's face mask.

Use and Benefits	Limitations
Supply of air is not time-restricted if a compressor is used.	Hose can be long, but not endless.
Positive pressure inside facepiece.	
Wearer is not burdened with cylinder.	

- **Self-Contained Apparatus**

Breathable air is supplied from a pressurised cylinder worn by the user. This type of BA gives the wearer complete freedom of movement, but it is the most heavy and bulky type. The air cylinder does have a limited capacity.

Use and Benefits	Limitations
Complete freedom of movement.	Supply of air is time-restricted.
Positive pressure inside facepiece.	Equipment is bulky and heavy.
	More technical training is required.

Selection, Use and Maintenance

RPE must be selected carefully to ensure that it is suitable.

> **TOPIC FOCUS**
>
> Key factors in the selection of RPE:
>
> - Contaminant concentration and its hazardous nature (e.g. harmful, toxic).
> - Physical form of the substance (e.g. dust, gas, vapour).
> - Level of protection offered by the RPE.
> - Presence or absence of normal oxygen concentrations.
> - Duration of time that it must be worn.
> - Compatibility with other PPE that must be worn.
> - Shape of the user's face and influences on fit.
> - Facial hair might interfere with an effective seal.
> - Physical requirements of the job, e.g. the need to move freely.
> - Physical fitness of the wearer.
> - Work environment, e.g. temperature and humidity.

The level of protection offered by an item of RPE is usually expressed as the Assigned Protection Factor (APF). This is a measure of how well the RPE keeps out the contaminant and is given by the formula:

$$APF = \frac{\text{Concentration of contaminant in workplace}}{\text{Concentration of contaminant in facepiece}}$$

Any RPE selected must meet the relevant standards (e.g. 'CE' marked).

Users of RPE should receive appropriate information, instruction and training. In particular, they should:

- Understand how to fit the RPE.
- Have a face-fit test to ensure suitability and fit (Reg. 7 of **COSHH**).
- Know:
    - How to test the item during use to ensure it is working effectively.
    - The limitations of the item.
    - Any cleaning requirements.
    - Any maintenance requirements (e.g. how to change the filter).

Maintenance and cleaning of RPE must be carried out in accordance with the manufacturers' instructions and any legal requirements (e.g. **COSHH** requirements for keeping a record of inspections, replacement parts, etc.). This should include the need to repair or replace worn or damaged items. Maintenance should be carried out by trained, competent personnel.

Other Protective Equipment and Clothing

There are other types of PPE used to protect from exposure to hazardous substances.

Gloves

Gloves (short cuffs) and gauntlets (long cuffs) can give protection against:

- Chemicals such as acids, alkalis, solvents, oils (e.g. corrosive cement).
- Biological agents such as blood viruses and body fluids.
- Physical agents such as contaminated dusts, and cuts from contaminated blades or syringes.
- Water - even uncontaminated water can soften and damage the skin of someone exposed for long periods.

For protection against chemicals, it is important to ensure the gloves are of the right material impervious to the chemical.

Overalls

'Ordinary' overalls are not regarded as PPE, but 'workwear' - in that they are not commonly intended to do more than keep a person (and their clothing) clean. However, even overalls can offer some level of protection against everyday construction contaminants such as soil, clay, oils and grease.

There are, however, items of PPE intended to protect the construction worker from hazards:

- Flame-retardant overalls.
- Chemical-resistant overalls - protect from acids, alkalis, etc.
- Disposable coveralls (hooded) - worn in asbestos removal and impervious to the passage of extremely fine fibres.
- Aprons - prevent spills and splashes from soaking into normal workwear and the skin.

Eye Protection

Four different types of eye protection are common in construction activities:

- **Spectacles**
    - Offer a degree of front and side protection, but do not completely encase the eyes.
    - Mainly for impact protection from flying objects and debris.
- **Safety Goggles**
    - Completely encase the eyes with protection from impact, chemical gas, liquid splashes and molten metal.

12.3 Control Measures

- **Face Shields**
  - Cover the eyes and face, but do not enclose them.
  - Limited protection from impact and splashes.
- **Hoods and Visors**
  - Offer all-round enclosed protection, especially from liquid splashes to the face.

Safety Helmets

Protection from falling or moving objects on a construction site is required, regardless of exposure to hazardous substances.

Safety Footwear

This not only offers protection to the toes, but some may be chemical-resistant where exposure to spillages or contaminated ground may occur. On construction sites, **CDM** specifically recognises that sharp objects where employees walk and stand are a significant risk. Midsole protection should be provided on construction safety footwear to guard against nails projecting from boards and other objects. To meet this standard, the footwear must be able to resist penetration using either a steel, aluminium or Kevlar insert to the midsole.

Personal Hygiene and Protection Regimes

Personal hygiene goes a long way to preventing absorption of hazardous substances into the body, by preventing contact in the simplest of situations. Many chemical and biological agents get into the body from the skin of the hands, into the mouth and eyes from cross-contamination. Likewise, food, drink and cigarettes all offer the same opportunity.

Good hygiene means:

- **Hand-washing** when leaving the work area, and always before eating, drinking or smoking.
- **Careful removal** and disposal of potentially contaminated items of PPE to prevent cross-contamination to normal clothes and the skin.
- **Prohibition** of eating, drinking and smoking in work areas.

All construction sites must have adequate welfare facilities (water, soap and a means for drying). With more serious hazards, showers and nail brushes may be required. Barrier creams may also prove useful.

Facilities should be provided to:

- Change and store clothing and PPE.
- Store, prepare and consume food and drinks.

Hand-washing can prevent the transfer and ingestion of hazardous substances

In some situations, **vaccinations** may protect workers from biological agents:

- Vaccination against hepatitis B is often offered to first aiders.
- Those working near water may gain some protection from immunisation against Weil's disease.

Issues to consider before embarking on a vaccination programme:

- Worker consent must be obtained.
- Vaccination does not always grant immunity.
- Vaccination can give workers a false sense of security.

In most cases, vaccination should only be offered when indicated by law or codes of practice.

Health Surveillance and Biological Monitoring

Health surveillance is a system of ongoing health checks and often involves carrying out some form of medical examination or test on employees who are exposed to substances such as solvents, fumes, biological agents and other hazardous substances.

Health surveillance is important to enable early detection of ill-health effects or diseases, and also helps employers to evaluate their control measures and to educate employees. The risk assessment will indicate where health surveillance may be needed, **but it is required where**:

- there is an adverse health effect or disease linked to a workplace exposure; **and**
- it is likely that the health effect or disease may occur; **and**
- there are valid techniques for detecting early signs of the health effect or disease; **and**
- the techniques don't themselves pose a risk to employees.

MORE...

There is a range of industry-specific guidance on health surveillance at:

www.hse.gov.uk/health-surveillance

There are two types of health surveillance commonly carried out:

- **Health monitoring** - where the individual is examined for symptoms and signs of disease that might be associated with the agent in question. For example, those working in the dustiest areas of a site or in cement production may have lung-function tests (spirometry) to check for respiratory disorders.
- **Biological monitoring** - where a blood, urine or breath sample is taken and analysed for the presence of the agent itself or its breakdown products. For example, those working with lead processes might have blood samples taken to check for cumulative levels of lead in the blood.

When necessary, health surveillance should be conducted on first employment, to establish a 'baseline', and then periodically. It can also be done when a person leaves employment as a final check. The need for health surveillance is usually subject to regulations, ACoP and guidance.

Similar health checks may be required for those exposed to noise, vibration, etc., and are covered in the appropriate section of this course.

Additional Controls for Carcinogens, Asthmagens and Mutagens

COSHH require that exposure to substances that can cause cancer, asthma or damage to genes that can be passed from one generation to another should be prevented. If this is not possible, specific measures should be adopted, including:

- Total enclosure of the process and handling systems.
- Prohibition of eating, drinking and smoking in contaminated areas.
- Regular cleaning of floors, walls and other surfaces.
- Designation of areas that may be contaminated with warning signs.
- Safe storage, handling and disposal.

12.3 Control Measures

STUDY QUESTIONS

17. What principles of control are illustrated by the following measures?
    (a) Using granulated pottery glazes instead of powders.
    (b) Vacuum cleaning rather than sweeping with a broom.
    (c) Job rotation.
    (d) Using water-based adhesives rather than solvent-based ones.
18. What is the difference between local exhaust ventilation and dilution ventilation?
19. What are dead areas, and why are they a problem for dilution ventilation?
20. List four main types of respirator and the three main types of breathing apparatus.
21. What are the key criteria in the selection of the appropriate respirator to use?
22. What is the main purpose of routine health surveillance?

(Suggested Answers are at the end.)

Specific Agents

IN THIS SECTION...

- A range of construction activities can create a risk of exposure to hazardous substances that can damage the lungs when breathed in and cause lung disease if not properly controlled.
- Specific hazardous agents may be encountered in construction work that can cause ill health to workers exposed, such as blood-borne viruses, carbon monoxide, cement, *Legionella*, *Leptospira*, silica, and tetanus.
- Asbestos, although now banned, can still be found in older buildings. Systems must be in place to:
    - Identify the presence of asbestos.
    - Procedure for discovery.
    - Protect workers.
    - Remove and dispose of asbestos.
    - Protective equipment.
    - Safe disposal.

The Prevalence of Occupational Lung Disease Among Construction Workers

Construction workers have a higher risk of developing occupational lung diseases than others because many common construction tasks can create high dust levels. Repeated exposure to dust can lead to a range of lung diseases like cancer, asthma, silicosis, and chronic obstructive pulmonary disease (COPD).

Construction dust is a general term used to describe different dusts that may be found on a construction site. There are three main types:

- Silica dust – created when working on materials like concrete, mortar and sandstone that contain silica. (The section below contains more information on silica).
- Wood dust – created when working on wood and other wood-based products like MDF. (See the section below for more information on wood dust).
- Other 'general' dust – created when working on materials that contain little or no silica, such as gypsum (found in plasterboard), limestone and marble.

Common construction tasks that can produce high levels of dust include:

- Cutting paving blocks, kerbs and flags.
- Chasing concrete and raking mortar.
- Scabbling or grinding.
- Cutting and sanding wood.
- Sanding taped and covered plasterboard joints.
- Soft strip demolition.
- Dry sweeping.

Many construction workers are exposed daily but can frequently change employers and work for short durations on many sites, and, because the amounts breathed in seem small or insignificant, respiratory risks are often overlooked. However, dust can build up in the lungs and cause harm gradually, it can take years before symptoms of ill-health become apparent. These diseases cause permanent disability and early death.

12.4 Specific Agents

COSHH covers activities which have the potential for construction workers to be exposed to construction dust. You need to:

- Assess the risks linked to the work and the materials, looking at the:
  - Task – the more energy the work involves, the bigger the risk.
  - Work area – the more enclosed a space, the more the dust will build up.
  - Time – the longer the work takes, the more dust there will be.
  - Frequency – regularly doing the same work, day after day, increases the risks.
- Control the risks, using the following control measures:
  - Stop or reduce the dust.
  - Control the dust – by using water or on-tool extraction.
  - Respiratory protective equipment (RPE) that is adequate for the amount and type of dust, suitable for the work, compatible with other PPE, fits the user, and is worn correctly.
- Review the controls to ensure they are working correctly.

To reduce the incidence of occupational lung diseases in workers within the construction industry, there needs to be a proactive raising of awareness of the risks of exposure, via inhalation, to hazardous dusts. Workers' knowledge needs to be improved – anyone who breathes in dust needs to know the damage that can be done to the lungs and airways. Good practices to minimise the risks should be promoted and encouraged.

Health Risks, Controls and Likely Workplace Activities/Locations Where They Can be Found

The following are some chemical and biological agents commonly encountered in construction activities, with a description of their health effects and the workplace circumstances in which they might be present.

Blood-Borne Viruses

Blood-borne viruses are carried in the bloodstream of an infected person, but are not easily transmitted to others unless their blood comes into contact with the broken skin of another person. Such viruses include the Human Immunodeficiency Virus (HIV) which causes Acquired Immunodeficiency Syndrome (AIDS) and Serum Hepatitis (hepatitis B).

Exposure to blood-borne viruses usually comes from contact with blood from injured persons being treated by a first aider at work, or from unintentional contact with discarded items such as needles ('sharps injuries' and 'needlestick' injuries).

Typical controls include:

- The use of gloves and eye protection when handling potentially contaminated material.
- Correct collection and disposal of potentially contaminated material.
- Preventing needlestick injuries by correct collection and disposal of sharps in a sharps container.
- Vaccination where appropriate.
- Procedures to deal with accidental exposures (e.g. needlestick injury).

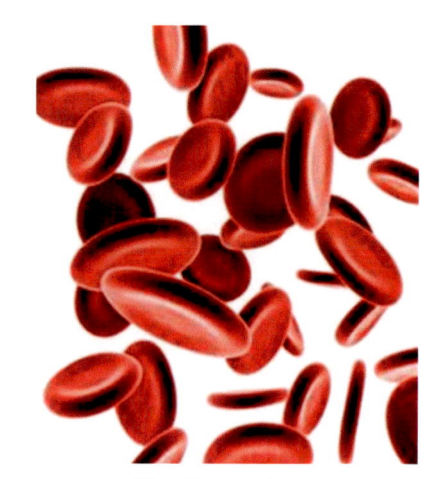

Blood-borne viruses

Carbon Monoxide

Carbon monoxide (CO) is a colourless, odourless, tasteless gas. It is found in combustion gases such as coal gas, car exhaust, producer gas, blast-furnace gas and water gas.

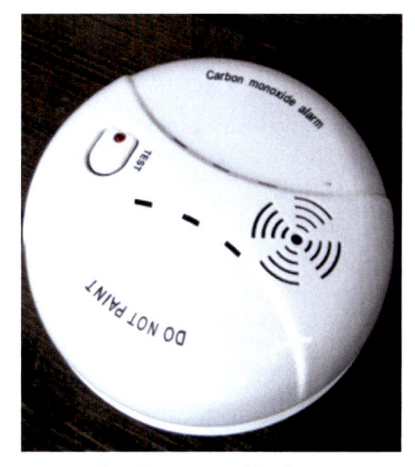

CO is toxic. It combines with haemoglobin in the blood, impairing the transportation of oxygen. Concentrations above 5% cause immediate loss of consciousness, but far more people are killed by exposure to much lower concentrations over an extended period, typically when a gas-fired heater is used in a poorly ventilated room, e.g. when used to heat a poorly ventilated site office.

Typical controls include:

- Restrict work on gas systems to competent engineers only.
- Maintenance and testing of boilers and flues.
- Good workplace ventilation.
- LEV for vehicle exhausts in workshops.
- Care in the siting of plant run by internal combustion engines.
- Carbon monoxide alarms.
- Confined space entry controls.

A carbon monoxide alarm can save lives

Cement

Cement is widely used in construction, e.g. mortar, plaster and concrete, and presents a hazard to health in a number of ways, mainly by skin contact and inhalation of dust.

Contact with wet cement can cause both dermatitis and burns:

- Cement is capable of causing dermatitis by irritancy and allergy.
- Both irritant and allergic dermatitis can affect a person at the same time.
- Wet cement can cause burns due to its alkalinity.
- Serious chemical burns to the eyes can also be caused following a splash of cement.

Cutting concrete creates high levels of cement and silica dust

Typical controls include:

- Eliminating or reducing exposure.
- Use of work clothing, and PPE such as gloves, dust masks and eye protection.
- Removal of contaminated clothing.
- Good hygiene and washing on skin contact.
- Health surveillance of skin condition to control chrome burns and dermatitis.

12.4 Specific Agents

Legionella

Legionnaires' disease is caused by the water-loving soil bacterium *Legionella pneumophila*, as is Pontiac fever, a shorter, more feverish illness, without the complications of pneumonia. Legionellosis is the generic term used to cover Legionnaires' disease and Pontiac fever.

The bacteria are hazardous when inhaled in a droplet form mixed with water. The most common sources for outbreaks of the disease are outdoor cooling towers associated with air-conditioning systems. Water containing bacteria is sprayed inside the towers to cool, and mists drift out of the top and are inhaled by passers-by.

Legionnaires' disease developing to the pneumonia stage can often prove fatal, especially for the elderly, infirm or immuno-suppressed, and for anyone if not diagnosed early.

Typical controls include:

- **Management Controls**
    - Assessment of the risk from *Legionella*.
    - A written control scheme (see below).
    - Appointment of a 'responsible person' to carry out risk assessment, and manage and implement the written scheme of controls.
    - Review of control measures.

    There are also duties placed on those involved in the supply of water systems.

- **Practical Controls**
    - Avoid water temperatures between 20°C and 45°C and conditions that favour bacterial growth.
    - Avoid water stagnation which can encourage biofilm growth.
    - Avoid using material that can harbour bacteria and provide them with nutrients.
    - Control the release of water spray.
    - Keep water, storage systems and equipment clean.
    - Use water (chemical) treatments where necessary.
    - Carry out water sampling and analysis.
    - Ensure correct and safe operation and maintenance of water systems.

- **Written Scheme of Controls**

    With reference to risk assessment, the written scheme should show:
    - An up-to-date plan of the plant or water system layout, including parts temporarily out of use.
    - A description of the correct and safe operation of the system.
    - What checks are required to ensure the written scheme is effective.
    - The frequency of such checks.
    - What remedial actions are to be taken if the written scheme is ineffective.

    MORE...

    ACoP L8, *Legionnaires' disease - The control of legionella bacteria in water systems*, available at:

    www.hse.gov.uk/pubns/books/l8.htm

Specific Agents | 12.4

Leptospira

DEFINITION

ZOONOTIC DISEASE (ZOONOSES)

Diseases which originate in animals and can be passed to humans (e.g. rabies).

Leptospirosis (Weil's disease) is a zoonotic disease caused by *Leptospira* bacteria. Rats are the primary cause of the disease (from their urine deposits) but it can also be found in mice, cattle and horses. The primary routes of infection are by swallowing contaminated water or food, and through cuts and grazes. Persons at risk include canal workers, sewer workers, rat catchers and agricultural workers.

Leptospirosis starts with 'flu-like symptoms - fever, headache and muscle pain - then progresses to a more serious jaundice-like phase. At this stage it can cause liver damage. It can be immunised against and, if diagnosed early, can be successfully treated. If left, it can be fatal.

Rats are the primary carrier of the Leptospira bacteria

Typical controls include:

- Preventing rat infestation by good housekeeping and pest control.
- Good personal hygiene (e.g. hand-washing).
- PPE (especially gloves).
- Covering cuts and grazes.
- Issuing workers with an 'at risk' card to be shown to doctors to assist early diagnosis.

Silica

Silica is a compound present in many rocks and stones, particularly sandstone, quartz and slate, and found in ceramics (e.g. clay pipes) and cement. It is hazardous when inhaled as a dust and can cause numerous chest and respiratory tract diseases. The most common is where silica is deposited deep in the lungs, causing scar tissue to form (silicosis) very similar to asbestosis.

Typical controls include:

- Prevention of exposure by use of alternative work methods.
- Dust suppression by use of water sprays or jets.
- Local exhaust ventilation.
- Respiratory protective equipment.
- Health surveillance (spirometry and chest X-ray).

Wood Dust

Wood is classified in two broad categories - hardwood and softwood. The main activities causing problems arise from sawing, routing, sanding and turning. Medium Density Fibreboard (MDF) is a composite of and resin and particularly causes asthma.

The smaller the particle size and the greater the amount of dust produced, the greater the risk of ill health.

Health problems associated with exposure to include:

- Skin disorders.
- Obstruction in the nose, rhinitis and asthma.
- A rare type of nasal cancer.

12.4 Specific Agents

- Wheezing, coughing and breathlessness.
- Stomach disorders due to ingestion.
- Eyes - soreness, watering, conjunctivitis.

Hardwood and softs both have a WEL of 5mg/m^3 which must not be exceeded. Because some s are asthmagens, exposure must be reduced to as low as is reasonably practicable.

Key controls:

- LEV must be provided (maintained and inspected) to extract dust at woodworking machines.
- should be vacuum-collected rather than swept or blown with an airline. Such systems should be suitable and have a High-Energy Particle Arrestor (HEPA) filter.
- RPE, as well as LEV, should be used for particularly dusty tasks.
- Health surveillance may be appropriate for respiratory disorders (asthma in particular).

Tetanus

The organisms causing tetanus (lockjaw) are widespread, usually found in vegetation, contaminated soil and animal excretions. They can gain access to the body through cuts, wounds, splinters, etc. Symptoms include stiffness in the muscles, a stiffening of the jaw until it is in a locked position, and breathing problems. There is a mortality rate of approximately 10%.

Typical controls include:

- Immunisation of workers to help to prevent the disease. Construction workers are susceptible when working on new sites or where there has been any agricultural activity taking place. An immunisation programme should be encouraged for any such workers.
- Use of strong gloves when handling materials that could be contaminated.
- Good hygiene and hand-washing.

Health Risks from and Controls for Working with Asbestos

This is a mineral of fibrous nature, capable of causing asbestosis, lung cancer and mesothelioma from inhalation. It has exceptional heat insulating qualities. Its use is banned in the UK, but it may still be found in older properties, where it is required to be managed. The biggest danger in construction is unwittingly locating asbestos during demolition or refurbishment work involving installation and even minor repairs.

We will look at methods for controlling exposure to asbestos later in this element.

Asbestos in older buildings is required by law to be managed

Specific Agents | 12.4

> **TOPIC FOCUS**
>
> ### Regulations Governing Asbestos Work
>
> The main regulations governing all work with asbestos on workplace premises are the **Control of Asbestos Regulations 2012**, which:
>
> - Cover everyone who is liable to be exposed to asbestos.
> - Maintain an explicit duty on employers and occupiers in non-domestic premises to manage asbestos.
>
> The regulations continue the prohibition on the importation, supply and use of all forms of new asbestos, and allow existing Asbestos-Containing Materials (ACMs) in good condition to be left in place, so long as their condition is monitored and managed to ensure they are not disturbed. The prohibition remains on the use of all second-hand asbestos products.
>
> Work with asbestos cement and similar ACMs remains unlicensed, but under certain conditions notification (not licensing) is still required and medical surveillance necessary as for licensed work.

During construction (in particular demolition), ACMs may be found in roofs (sprayed coatings and asbestos cement sheets), walls (insulation and fire protection), floors and ceilings (tiles), rainwater drainage (asbestos cement gutters, drainpipes and fountain heads) and as pipe lagging.

Three respiratory diseases are associated with asbestos exposure:

- **Asbestosis** - asbestos fibres lodge deep in the lungs and cause scar tissue to form. Extensive scarring leads to breathing difficulties and increases the risk of cancer - can be fatal.
- **Lung cancer** - asbestos fibres in the lung trigger the development of cancerous growths in lung tissue - often fatal.
- **Mesothelioma** - asbestos fibres migrate through the lung tissue and into the cavities around the lung, triggering the development of cancerous growths in the lining tissue around the lungs, the heart and the lining of the abdomen.

> **MORE...**
>
> Further information on asbestos can be found at:
>
> www.hse.gov.uk/asbestos/index.htm

The effects of asbestos have a long latency period - it will be a long time after exposure has occurred before symptoms are apparent (10-15 years for asbestosis; 30-40 years for mesothelioma).

Asbestos is now banned in the UK, but it may still be present in older buildings and encountered during maintenance work or demolition.

Duty to Manage Asbestos

Asbestos Identification

Asbestos or ACMs are not easily recognised, so laboratory testing of samples is often required, and it is even more difficult when ACMs are concealed by decorations or coatings. ACMs may be present if the building was constructed or refurbished before blue and brown asbestos were banned in 1985, or asbestos cement in 1999.

The three main types of asbestos that have been used commercially are:

- Crocidolite (blue).
- Amosite (brown).
- Chrysotile (white).

12.4 Specific Agents

The type of asbestos cannot be identified just by its colour, but requires microscopic examination in a laboratory.

Some products have one type of asbestos in them while others have mixtures of two or more. All types are dangerous, but blue and brown asbestos are known to be more dangerous than white.

Types of Survey and Who Can Undertake Them

In order to ascertain if asbestos is present in a building, a survey can be carried out.

If there is doubt as to whether asbestos is present, it should be presumed that it is present and also that it is not restricted to white asbestos, and so the regulations would apply accordingly.

Sometimes, samples have to be taken for analysis. Conducting a survey can be done by in-house competent people but, in many instances, must be done by an external **competent surveyor**. This kind of survey is referred to as a '**management survey**'.

A building survey is carried out to determine the presence of asbestos

> **TOPIC FOCUS**
>
> To establish if asbestos is present in a building, a **survey** should be carried out. The types of surveys are:
>
> - **Management Survey** - to manage ACMs during the normal occupation and use of the premises. In simple and straightforward premises, the dutyholder can do this, otherwise a surveyor is needed. This will establish that:
>   - Nobody is harmed by the asbestos remaining in the premises or equipment.
>   - The ACMs remain in good condition.
>   - Nobody disturbs them accidentally.
>
>   This may require minor intrusion and asbestos disturbance to locate ACMs that could be disturbed or damaged by normal activities, maintenance or equipment installation. It guides the occupier in prioritising remedial work that may be needed.
>
> - **Refurbishment/Demolition Survey** - required where the premises or part of it needs upgrading, refurbishment or demolition. This would not need a record of the ACM condition, and should be done by a surveyor. This will establish that:
>   - Nobody is harmed by work on ACMs in the premises or equipment.
>   - Such work will be done by the right contractor in the right way.
>
>   This involves destructive inspection and asbestos disturbance to locate ACMs before structural work starts. The area must be vacated and certified 'fit for reoccupation' after the survey.

Where it Can be Located

Some examples of asbestos use include:

- **Insulation board** - contains around 20-45% asbestos - used for fire protection; heat and sound insulation; in ducts; in-fill panels; ceiling tiles; wall linings; bath panels and partitions; and fire doors.
- **Pipe lagging** - contains 55-100% asbestos - used for thermal insulation on boilers and pipes.
- **Fire blankets** - used in homes and commercial catering kitchens.
- **Floor tiles** - very similar in appearance to ordinary vinyl or plastic tiles.

- **Sprayed coatings/loose fill** - inside roofs, lofts, etc.
- **Rope and gaskets** - used as seals around jointed pipes and in joints in boilers.
- **Roof felt** - rolls of felt laid on roofs; roof tiles.
- **Decorative paints and plasters** - lining walls, around and beneath staircases, etc.; 'artex' ceiling coatings.

Asbestos cement products include:

- Corrugated roof sheets.
- Rainwater goods (fountain heads, guttering, drain pipes, etc.).
- Cold water tanks and toilet cisterns.

Procedure for Asbestos Discovery During Construction

A procedure must be in place covering the actions to take on discovering asbestos in unknown locations. This will include stopping work and immediately informing the site supervisor. They should arrange for the area to be sealed off until a formal survey can be carried out.

Requirements if People Are Accidentally Exposed to Asbestos Materials

Requirements:

- Stop work immediately.
- Prevent anyone entering the area.
- Arrangements should be made to contain the asbestos - seal the area.
- Put up warning signs - 'possible asbestos contamination'.
- Inform the site supervisor immediately.
- If contaminated, all clothing, equipment, etc. should be decontaminated and disposed of as hazardous waste.
- Undress, shower, wash hair; put on clean clothes.
- Contact a specialist surveyor or asbestos removal contractor.

Requirements for Removal

Provided asbestos is contained and left undisturbed it can be retained, sealed and managed.

Where asbestos is to be removed, then a number of controls apply covering notification, licensing of operators and procedures.

Non-Licensed

Some work can be carried out without formal licensing (although strict controls will still be required). Such **non-licensed work** is where:

- The work is sporadic and of low intensity.
- Risk assessment shows the exposure will not exceed any limits.
- The work involves:
    - Short, non-continuous maintenance activities.
    - The removal of materials in which asbestos fibres are firmly linked to a matrix (e.g. asbestos cement sheet).
    - Encapsulation or sealing of ACMs in good condition.
    - Air monitoring and sampling.

12.4 Specific Agents

Licensed

Must be performed by a licensed contractor, and refers to activities which involve "*significant hazard, risk or public concern.*" Licenses can be granted by the HSE, who state that all work involving this type of activity must be brought to the attention of the "*appropriate enforcing authority*".

Notification of Licensed Work

If the work is licensable, at least 14 days' notice to the enforcing authority (or such shorter time as the enforcing authority may agree) is to be given. The Notification (as per Schedule 1 of the **Control of Asbestos Regulations 2012**) should contain:

- The name of the notifier and the address and telephone number of their usual place of business.
- The location of the work site.
- A description of the type of asbestos to be removed or handled (crocidolite, amosite, chrysotile, etc.).
- The maximum quantity of asbestos to be held at any one time on the premises at which the work is to take place.
- The activities or processes involved.
- The number of workers involved.
- Any measures taken to limit the exposure of employees to asbestos.
- The date of commencement of the work activity and its expected duration.

Even where work is technically 'non-licensable' work, the relevant authorities must still be notified of the work if:

- it is carried out for more than two hours in any seven days; or
- it takes more than one hour to complete.

Plan of Work

Work with asbestos must not start without a written plan detailing how the work is to be carried out. The plan must be kept at the premises where the work is being carried out for the full duration of the work. For final demolition or major refurbishment, the plan will usually require that asbestos is removed before any other major works begin.

The plan will specify what control measures are required for managing the risk, including:

- Monitoring the condition of any asbestos or ACMs.
- Ensuring asbestos or ACMs are maintained or safely removed.
- Providing information about the location and condition of any asbestos or ACMs to anyone liable to disturb it/them.
- Making this information available to the emergency services.

The measures specified in the plan must be implemented and recorded.

The plan is to be reviewed and revised:

- At regular intervals.
- If there is reason to suspect that it is no longer valid.
- If there has been a significant change in the premises to which the plan relates.

Specific Agents | 12.4

TOPIC FOCUS

Typical control measures for removing asbestos include:

- Restrict access to the area.
- Enclose the work area and keep it under negative pressure, testing the sealed area for leaks.
- Provide appropriate PPE (coveralls, respirators, etc.) and a decontamination unit.
- Ensure removal operatives are suitably trained.
- Use controlled wet removal methods (e.g. water injection, damping down the surface to be worked on). Dry removal processes are unacceptable.
- Use a wrap-and-cut method or glove bag technique (a method of removing asbestos from pipes, ducts, valves, joints and other non-planar surfaces).
- Where appropriate, use measures which control the fibres at source, e.g. by using vacuuming equipment directly attached to tools. Failing this, a second employee can use a hand-held vacuum right next to the source emitting the fibres (known as 'shadow vacuuming').
- Thoroughly clean the area and obtain a clean air certificate after a successful air test upon completion of the work.

A clearance certificate for reoccupation may only be issued by a body accredited to do so. At the moment, such accreditation can only be provided by the United Kingdom Accreditation Service (UKAS).

Respiratory Equipment

Suitable RPE should always be provided where exposure can be above the control limits.

The RPE provided must be marked with a 'CE' symbol and matched to:

- The exposure concentrations (expected or measured).
- The job.
- The wearer.
- Factors related to the working environment.

RPE must be examined and tested at suitable intervals by a competent person, and a suitable record kept for five years. Respirator testing involves daily checks, monthly checks, and full performance checks every six months. Operator checks would involve fit testing to see that the correct size and model are used to provide an adequate face seal.

Protective Clothing

- **Overalls**
    - **Only** wear **disposable** (hooded) overalls - type 5 (**BS EN ISO 13982-1:2004+A1:2010**) are suitable (cotton not recommended).
    - Wear waterproof overalls for outdoor work.
    - A few tips include:
        - Wear one size too big to avoid splitting the seams.
        - If the cuffs are loose, seal with tape.
        - Avoid long-sleeved shirts - they are difficult to cover properly.
        - Wear the overall legs over footwear, tucking them in lets dust into footwear.
        - Wear the hood over the RPE straps.
        - Dispose of used overalls as asbestos waste.
    - **Caution - never take used overalls home**.

12.4 Specific Agents

- **Gloves**
    - If worn, use single-use disposable gloves.
    - If latex, choose 'low-protein powder' gloves.
    - Dispose of as asbestos waste.
- **Footwear**
    - Boots are preferable to disposable overshoes which may cause risk of slipping.
    - **Caution - never use laced boots** - they have lace holes to catch asbestos fibres and are difficult to clean.

Training

Anyone removing asbestos must have training that includes:

- Properties of asbestos, its effects on health, including its interaction with smoking.
- The types of products or materials likely to contain asbestos.
- The operations which could result in asbestos exposure.
- Safe work practices, preventive control measures, and protective equipment.
- The purpose, choice, limitations, proper use and maintenance of RPE.
- Emergency procedures.
- Hygiene requirements.
- Decontamination procedures.
- Waste handling procedures.
- Medical examination requirements.
- The control limit and the need for air monitoring.

Employees should be made aware of the significant findings of the risk assessment, and the results of any air monitoring carried out, with an explanation of the findings.

Air Monitoring

Sampling for asbestos in the air should be carried out by trained staff, in three situations:

- **Compliance sampling** - within control or action limits.
- **Background sampling** - before starting work (i.e. removal).
- **Clearance sampling** - after removal and cleaning the area.

Medical Surveillance

Requirements for surveillance for persons who are exposed to asbestos:

- A health record is to be kept (for 40 years) and maintained.
- Surveillance for those carrying out licensable work with asbestos requires a doctor to carry out a medical examination every two years; and every three years for those carrying out non-licensable work requiring notification.
- A certificate should be issued to the employer and the employee and kept for four years.

Requirements For Disposal

Asbestos waste is hazardous waste if it contains more than 0.1% asbestos. The **Hazardous Waste (England and Wales) Regulations 2005** apply, and a waste consignment note is required:

- Double wrap and label the waste - standard practice is to use a red (UN) inner bag, and a clear outer bag with **Carriage of Dangerous Goods and Use of Transportable Pressure Equipment Regulations 2009** warnings and asbestos code visible.
- The waste must be carried in a sealed skip or vehicle with a segregated compartment for asbestos, easily cleanable and lockable.
- It must be transported by a licensed waste carrier and be taken to a licensed disposal site. The waste consignment note is to be kept for three years.

STUDY QUESTIONS

23. Identify six chemical and four biological agents commonly encountered in construction activities.
24. Identify two asphyxiant gases and outline their ill-health effects.
25. Identify three sources of organic solvents used in construction, and describe their ill-health effects.
26. Identify the controls used to avoid or reduce exposure to cement dust and wet cement.
27. Identify the three main types of asbestos.
28. What steps are to be taken if you discover asbestos on site?
29. What are the three air-monitoring sampling methods for asbestos, and when should they be carried out?

(Suggested Answers are at the end.)

Summary

This element has dealt with some of the hazards and controls relevant to chemical and biological health hazards in the construction environment.

In particular, this element has:

- Outlined the different forms of chemicals (liquids, gases, vapours, mists, fumes, fibres and dusts) and biological agents (fungi, bacteria and viruses).
- Identified the classification of chemicals (toxic, harmful, corrosive, irritant and carcinogenic) and the meaning of the terms 'acute' and 'chronic' when used to describe their effects.
- Identified the main routes of entry into the body (inhalation, ingestion, absorption through the skin and injection through the skin) and some of the body's defence mechanisms.
- Explained the factors to be considered when undertaking an assessment of the health risks from substances encountered in construction workplaces.
- Outlined the sources of information available about the substances, and the use of safety data sheets and product labels.
- Described some of the equipment that is used when undertaking basic surveys to assess concentrations of substances in the workplace (e.g. stain tube detectors).
- Explained the purpose and principles of occupational exposure limits and their relevance in short-term and long-term exposures.
- Outlined the control measures that should be used to reduce the risk of ill health from exposure to hazardous substances.
- Described the principles of good practice as regards to controlling exposure to hazardous substances: minimising emissions; taking into account routes of exposure; exposure control to be proportional to risk; choosing effective controls; using PPE; regular checks and reviews of controls in place; and provision of information and training. Control measures should not increase the overall risks.
- Described common measures to implement the principles of good practice: eliminate or substitute; change the process; reduce exposure time; enclose or segregate; LEV; dilution ventilation; RPE and PPE; personal hygiene; and health surveillance.
- Outlined the hazards, risks and controls associated with specific hazardous agents.
- Described the health risks and controls associated with asbestos and the duty to manage asbestos.

Exam Skills

Question

Scenario

You have come out of a meeting with the occupational health specialist who has concerns about silica dust amongst construction workers. During the meeting, the site manager has asked you for a briefing document on this subject.

Task: Chemical Hazards

Back in the office you decide to prepare a briefing document to be used at a toolbox talk on the subject of silica dust. In the briefing document you are to cover the following areas:

(a) Construction activities that may expose workers to silica dust.	**(4 marks)**
(b) Health risks associated with silica dust.	**(2 marks)**
(c) Control measures to reduce the risk from silica dust.	**(4 marks)**
	(Total: 10 marks)

Approaching the Question

Now think about the steps you would take to answer this question:

Step 1 The first step is to **read the scenario carefully**. Note the question is focussing on a particular chemical hazard, where it can be encountered the sort of harm and means of controlling the risk from it.

You decide to create a briefing document for a toolbox talk so you will need to structure your approach using the three headings given.

Step 2 Now look at the **task** - prepare some notes under the three headings "Construction activities exposing workers to silica dust", "Health risks associated with silica dust" and "Control measures to reduce the risk from silica dust".

Step 3 Next, consider the **marks** available. In this task, there are 4 marks available for the first part, 2 marks for the second part and 4 marks for the third part of the question. Tasks that are multi-part are often easier to answer because there are additional signposts in the question to keep you on track. In this task, you have to create a briefing document that is easy to understand, giving examples for each part can aid understanding. You will need to provide around 10 or more different pieces of information including examples for this task. The headings will allow you to keep your response separate – this will also help the examiner when marking.

Step 4 **Read the scenario and task again** to make sure you understand the requirements and ensure you have a clear understanding of the hazards associated with silica dust. (Re-read your study text if you need to.)

Step 5 The next stage is to **develop a plan** - there are various ways to do this. Creating a bullet point list could be one way.

Exam Skills

Suggested Answer Outline

Construction activities exposing workers to silica dust:

- Bricklayers cutting bricks with a saw.
- Pouring cement out of a bag.
- Crushing stone activities.
- Cutting paving slabs with a disc saw.

Health risks associated with silica dust:

- Chest and respiratory disorders.
- Silicosis.

Control measures to reduce the risk from silica dust:

- Prevention of exposure by use of alternative work methods.
- Dust suppression by use of water sprays.
- Respiratory protection equipment.
- Health surveillance.

Now have a go at the question yourself.

Example of How the Question Could be Answered

(a) Silica dust can be encountered in construction activities in several ways. First, through bricklayers cutting through clay bricks in order to reduce the size of the brick using a saw. During cement mixing operations where a cement bag is opened and poured into a cement mixer, causing the cement powder to become airborne in the air. During stone crushing activities where the stone dust becomes airborne. Finally, using a powered disc saw to cut through paving slabs or roofing materials such as slate.

(b) The health risks associated with silica dust can be various forms of chest or respiratory disorders as well as a disease called silicosis causing scar tissue in the lungs which is very similar to asbestosis.

(c) Control measures to reduce the risk of silica dust can be by prevention of exposure through the use of alternative work methods such as ordering ready mixed cement and by adopting hand cutting opposed to powered cutting. Wetting the area using dust suppression water sprays on site vehicle routes and where powered cutting is taking place. The use of RPE such as half face respirators. Monitoring workers over time through various health surveillance techniques such as spirometry and chest X-rays.

Reasons for Poor Marks Achieved by Exam Candidates

- Not following a structured approach for the briefing document; failing to provide information on the three subject areas.
- Not expanding the answer beyond a few words as opposed to giving a sentence of explanation.

Element 13

Physical and Psychological Health

Learning Objectives

Once you've read this element, you'll understand how to:

1. Outline the health effects associated with exposure to noise and appropriate control measures.

2. Outline the health effects associated with exposure to vibration and appropriate control measures.

3. Outline the health effects associated with ionising and non-ionising radiation and the appropriate control measures.

4. Outline the causes and effects of mental ill health at work and appropriate control measures.

5. Explain the hazards and appropriate control measures for violence at work.

6. Explain the hazards and appropriate control measures for substance misuse at work.

Contents

Noise — 13-3

Introduction to Noise	13-3
The Physical and Psychological Effects of Exposure to Noise	13-3
Commonly Used Terms in the Measurement of Sound	13-5
When Exposure Should be Assessed	13-5
Comparison of Measurements to Exposure Limits Established by Recognised Standards	13-6
Basic Noise Control Measures	13-7
Purpose, Application and Limitations of Personal Hearing Protection	13-8
Role of Health Surveillance	13-9

Vibration — 13-10

The Effects on the Body of Exposure to Vibration	13-10
When Exposure Should be Assessed	13-11
Comparison of Measurements to Exposure Limits Established by Recognised Standards	13-13
Basic Vibration Control Measures	13-14
Role of Health Surveillance	13-15

Radiation — 13-16

Differences Between Types of Radiation and their Health Effects	13-16
Typical Occupational Sources of Radiation	13-20
Basic Ways of Controlling Exposure to Radiation	13-21
Basic Radiation Protection Strategies	13-22
The Role of Monitoring and Health Surveillance	13-24

Mental Ill Health — 13-25

The Frequency and Extent of Mental Ill Health in the Construction Industry	13-25
Recognising Common Symptoms	13-25
Causes of and Controls for Mental Ill Health	13-27
Recognition That Most People with Mental Ill Health Can Continue to Work Effectively	13-29
Organisations That Provide Support	13-29

Violence at Work — 13-31

Introduction to Violence at Work	13-31
Types of Violence at Work	13-31
Effective Management of Violence at Work	13-33

Substance Abuse at Work — 13-35

Risks to Health and Safety from Substance Abuse at Work	13-35
Managing Substance Abuse at Work	13-35

Summary — 13-38

Exam Skills — 13-39

Noise

IN THIS SECTION...

- Exposure to excessive noise causes Noise-Induced Hearing Loss (NIHL) as well as other physical and psychological effects.
- Noise exposure standards are based on a worker's daily personal noise exposure. A continuous 85 dB(A) over an 8-hour shift is considered the upper limit.
- Noise exposure should be assessed by carrying out a survey using a sound level meter.
- Exposure to noise can be controlled by: isolation, absorption, insulation, damping and silencing.
- Two types of hearing protection are available - ear defenders (muffs) and ear plugs. Both are effective, but have limitations.
- Health surveillance will help to reduce the risks of noise-induced hearing loss and maintain adequate control measures.

Introduction to Noise

DEFINITION

NOISE

Any audible sound.

The **Control of Noise at Work Regulations 2005** provide the framework for addressing the problems of workplace noise, although they are almost exclusively concerned with the risk of damage to hearing.

The Physical and Psychological Effects of Exposure to Noise

In moderation, noise is harmless, but if it is too loud, it can permanently damage hearing. The danger depends on how loud the noise is and how long people are exposed to it.

- **Physical Effects of Noise**
    - Temporary reduction in hearing sensitivity as a result of short-duration exposure to excessively loud noise. This is known as **temporary threshold shift**.
    - Temporary ringing in the ears as a result of short-duration exposure to excessively loud noise.
    - Noise-Induced Hearing Loss (NIHL) - permanent loss of hearing as a result of repeated exposure to excessively loud noise. One cause of **permanent threshold shift** (along with age, infection, etc.)
    - Tinnitus - persistent ringing in the ears as a result of repeated exposure to excessively loud noise.

Exposure to loud noise can cause both temporary and permanent hearing loss

13.1 Noise

- **Psychological Effects of Noise**
  - Stress effects:
    - Caused by irritating nuisance/background noise.
    - Due to the inability to clearly hear or define spoken words, and determine sounds (from radio, television, street noises, etc.).
  - Safety effects:
    - Inability to hear:
      - hazards such as vehicles;
      - alarms and warning sirens; and
      - conversation and spoken instructions;

    as a result of background noise.
  - Difficulty concentrating and an increase in errors caused by nuisance/background noise.

The most serious effect is NIHL (industrial deafness). This is irreversible and caused by exposure over a long period of time to excessively loud noise. Surgery may reduce the damage in the case of acute injury to the eardrum, but there is no cure for NIHL.

Sound waves travel into the ear canal until they reach the eardrum. The eardrum passes the vibrations through the middle ear bones or ossicles into the inner ear. The inner ear is shaped like a snail and is called the cochlea. Inside the cochlea there are thousands of tiny hair cells which change these vibrations into electrical signals that are sent to the brain through the hearing nerve. The brain tells you that you are hearing a sound and what that sound is. Each hair cell has a small patch of stereocilia sticking up out of the top. Sound makes the stereocilia rock back and forth. If the sound is too loud, the stereocilia can be bent or broken. This will cause the hair cell to die and it can no longer send sound signals to the brain. In humans, once a hair cell dies, it will never grow back. The high frequency hair cells are the most easily damaged, so people with hearing loss from loud sounds often have problems hearing high pitched speech or music.

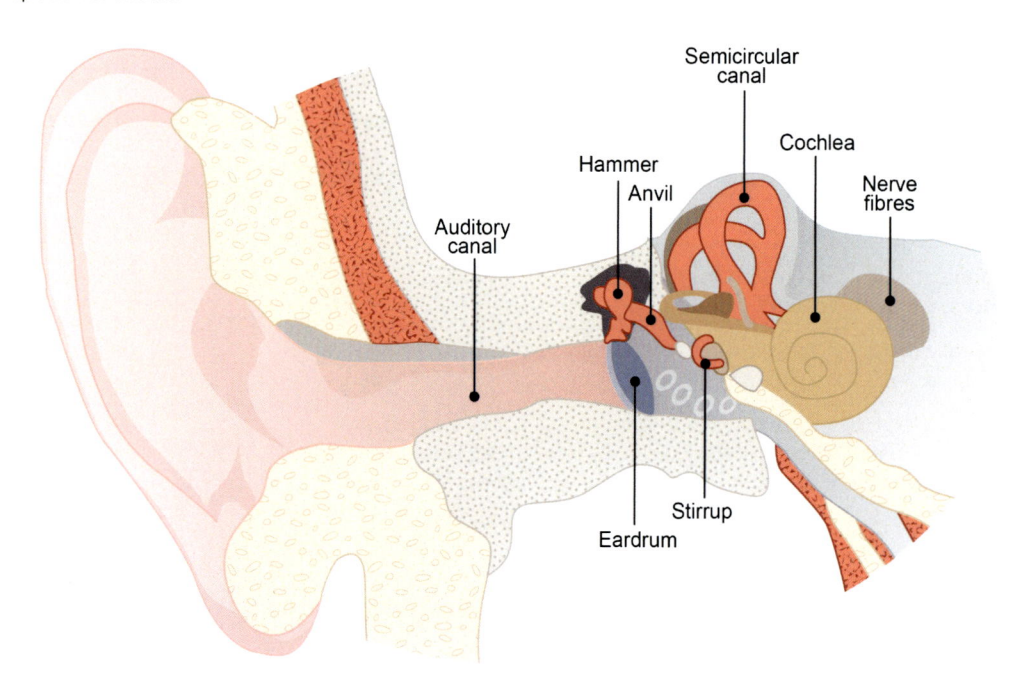

Diagram showing the internal parts of the ear

A quick guide to safe noise levels would be the 'Two-Metre Rule'. If normal conversation can't be heard two metres away from a person talking, the noise level is probably at or above safe limits.

One-off exposures to high noise levels (e.g. four hours' work in a high-noise area) cause a temporary loss of sensitivity (temporary threshold shift) due to disturbance of the cochlear hairs, and temporary ringing in the ears (tinnitus).

Repeated exposure over a number of years causes permanent damage in the cochlea (**permanent threshold shift**) and this is NIHL. The loss is progressive (it gets gradually worse over time), with earlier periods of damage being unable to repair themselves.

Commonly Used Terms in the Measurement of Sound

The following basic terminology is used in the measurement and assessment of sound and noise exposure in the workplace:

- **Sound pressure level** - a measure of the intensity of the sound pressure wave moving through the air. It is normally expressed as decibels (dB).
- **Intensity** - the measure of the flow of sound, and has *a level and a direction*. This flow is observed over a specific area, hence the units of sound intensity are W/m^2.
- **Frequency** - the number of sound pressure waves (e.g. contacting the eardrum) every second. It is measured in hertz (Hz). The human ear is sensitive to noise across a wide range of frequencies, from very low (bass) frequencies at 20 Hz to very high-pitch frequencies at 20,000 Hz.
- **Decibel (dB)** - the unit of sound pressure level (the 'loudness of the noise'). It is a logarithmic scale, which means that relatively small increases (or decreases) actually represent large increases (or decreases) in intensity. For instance, an increase of just 3 dB represents a doubling of sound intensity (i.e. 82 dB is twice as loud as 79 dB).
- **A-weighting** - applied during noise assessment to the decibel scale to give a sound pressure level expressed as **dB(A)**. This A-weighting takes into account the sensitivity of the human ear across a wide range of frequencies. In other words, it is the decibel value corrected for the human ear.
- **C-weighting** - applied during noise assessment to the decibel scale to give a sound pressure level expressed as **dB(C)**. This C-weighting gives a more accurate reading for impulse noise - such as single loud bangs that would not be properly recorded using the dB(A) scale.

Measurement in dB(A)	Sound
0	The faintest audible sounds
20-30	A quiet library
50-60	A conversation
65-75	A loud radio
90-100	A power drill
140	A jet aircraft taking off 25m away

When Exposure Should be Assessed

Noise exposure limits are set on the basis that the amount of damage done to the ear is directly related to the amount of energy absorbed by the inner ear. This is determined by two factors:

- The noise level (measured in dB(A)).
- The duration of exposure (in hours and minutes).

These two factors determine the '**dose**' of noise absorbed (a similar principle to hazardous substances and workplace exposure limits). It is therefore necessary, when undertaking a noise assessment, to measure a worker's **actual exposure** to noise (which will fluctuate) and then to calculate what the equivalent 8-hour exposure would be. This personal noise exposure is usually written as $L_{EP,d}$ (daily) or $L_{EP,w}$ (weekly).

13.1 Noise

Equal amounts of noise entering the ear cause the same effect, therefore a short exposure to a high level of noise is considered to cause comparable hearing damage to a longer exposure to a lower level of noise. So, where it may be considered 'safe' to be exposed to 85 dB(A) for 8 hours, exposure to 88 dB(A) would only be 'safe' for four hours. In both cases, the **dose** is the same. This is illustrated in the following table:

Exposure equivalents

Noise Level in the Workplace	Duration of Exposure	Noise 'Dose' ($L_{EP,d}$)
85 dB(A)	8 hours	85 dB(A)
88 dB(A)	4 hours	85 dB(A)
91 dB(A)	2 hours	85 dB(A)
94 dB(A)	1 hour	85 dB(A)
97 dB(A)	30 mins	85 dB(A)
100 dB(A)	15 mins	85 dB(A)
103 dB(A)	7.5 mins	85 dB(A)

As can be seen from the table above, if the sound level is doubled (represented by a 3 dB(A) increase) then the duration of exposure has to be halved for the total dose to remain the same. Noise, however, must not be allowed to exceed the exposure limit value which is explained shortly.

To take into account noise dose, the **Control of Noise at Work Regulations 2005** introduced **exposure action values** and **exposure limit values**. These are exposure values at which the employer is required to take particular steps to protect employees and others from the harmful effects of noise.

Comparison of Measurements to Exposure Limits Established by Recognised Standards

Lower Exposure Action Values (LEAVs)

These are:

- **80 dB(A)** - a daily or weekly (ambient) noise exposure not taking into account the effects of any hearing protection.
- **135 dB(C)** - a 'one-off' instantaneous noise (one impact) in the 8-hour day, not taking into account the effects of any hearing protection.

Where LEAVs may be exceeded, the employer must:

- Carry out a risk assessment and if this indicates a risk to health then carry out health surveillance.
- Provide information, instruction and training.
- Make hearing protection available.

Upper Exposure Action Values (UEAVs)

These are:

- **85 dB(A)** - as above, a daily, or weekly, personal exposure that does not take into account the effects of any hearing protection.
- **137 dB(C)** - as above, a 'peak sound pressure level' (a 'one-off' instantaneous noise) in the 8-hour day, not taking into account the effects of any hearing protection.

At or above UEAVs the employer must:

- Carry out a noise assessment.
- Reduce noise exposure to the lowest level reasonably practicable.

If noise levels are still above 85 dB(A), the employer must:

- Establish mandatory hearing protection zones.
- Provide information, instruction and training.
- Provide hearing protection and enforce its use.

Exposure Limit Values (ELVs)

These are:

- **87 dB(A)** - daily, or weekly, personal exposure that must not be exceeded. Hearing protection can be taken into account.
- **140 dB(C)** - a 'one-off' 'peak sound pressure level' during the working day. This must not be exceeded. Hearing protection can be taken into account.

If an ELV is exceeded, the employer must investigate the reason for the occurrence and identify and implement actions to ensure that it does not occur again.

Basic Noise Control Measures

In simple terms, noise exposure can be controlled in three ways:

- Reduce the noise at source.
- Interrupt the pathway from source to receiver.
- Protect the receiver.

Isolation

Noise is often transmitted from a machine into supporting structures by vibration (e.g. a compressor through the floor it stands on). Isolation involves separating the machine from noise-carrying structures using noise-absorbent mats or spring-mounts. This breaks the transmission pathway.

Absorption

Once noise has left its source, it can go directly through air to a receiver, or be reflected off hard surfaces (e.g. walls and ceilings). Absorption involves putting sound-absorbing materials in the way to absorb the sound waves before they can reach the receiver. Sound-absorbent coatings are often applied to walls to prevent noise reflection.

Insulation

Some noise sources can have an enclosure built around them. Any noise generated inside will remain there, without penetrating the walls into the rest of the workplace. Generators and compressors are often in their own noise-reduction enclosures.

Damping

Machine parts (especially metal surfaces) often resonate together, exaggerating the noise generated. Damping can change the resonance characteristics to prevent ringing noises. Change a part, stiffen it or coat one side with a sound absorbent.

Silencing

Any machine that produces exhaust air or gases (e.g. a diesel generator or pneumatic control valve) should be fitted with a silencer on the exhaust to suppress noise.

13.1 Noise

Purpose, Application and Limitations of Personal Hearing Protection

Types

Hearing protection prevents harmful noise from reaching the ear. Two principal types are available:

Ear Defenders (or Muffs)

These encase the outer ear in a cup with a sound-absorbent foam or gel-filled cushion to seal against the side of the head.

Advantages	Limitations
Easy to supervise and enforce use.	Uncomfortable when worn for long time.
Less chance of ear infections.	Incompatible with some other items worn (e.g. spectacles).
Higher level of protection possible; bone transmission is reduced.	Must be routinely inspected, cleaned and maintained.

Ear Plugs

These rigid, semi-rigid or foam plugs fit into the ear canal.

Advantages	Limitations
Cheap.	Difficult to see when fitted, so difficult to supervise and enforce.
Disposable.	
Often more comfortable to wear.	Risk of infection if dirty or if cross-contaminated when inserted.
Do not interfere with any other items worn.	Easy to lose or misplace.

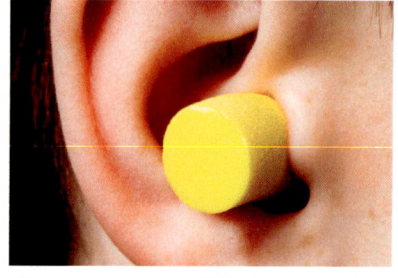

Ear plugs are more difficult to supervise and enforce

Selection

Hearing protection should be chosen taking into account the levels of noise the wearer will be exposed to and the strengths and limitations of each type. It must fit and be comfortable for the wearer.

As with all types of PPE, employers should ensure that only CE-marked hearing protection is purchased and used within their organisation.

Use

To be effective, hearing protection should be worn all the time the wearer is exposed. Much of the protection is lost when earmuffs or plugs are removed, even if only for a short time - removing the protection for only 15 minutes in an 8-hour shift can lose the wearer 80% or more of the protection. Not many users are aware of this fact. For enforcement and supervisory purposes, earmuffs can be seen more easily when worn.

Maintenance

Replacement of earmuffs and ear plugs is required at regular intervals, due to good personal hygiene practices or if the equipment is defective or damaged. Some parts on earmuffs can be changed without replacing the whole thing (e.g. ear cushions or absorbent pads).

Attenuation Factors

Noise received at the ear must be kept below any exposure limits, so information is required on the:

- Noise characteristics of the workplace (from the survey).
- Attenuation characteristics of the available hearing protection (from the manufacturers) - how much noise they remove.

Most forms of hearing protection give better protection against high-frequency noises - the high-pitched noises. The lower frequency rumbles from machinery are harder to reduce.

There is a general limitation on the level of noise reduction that can be achieved, depending on the quality and type of the ear protection. Ear defenders, because the cup covers the ear and rests on the side of the face, give better attenuation than plugs, as less noise can be transmitted through the bones of the skull.

Protection will be reduced by various factors which reduce the effectiveness of the seal between the ear and the earmuff or plug, e.g. as a result of long hair, thick spectacle frames or jewellery, incorrect fitting of plugs or the wearing of safety helmets or face-shields.

A range of factors can reduce the protection offered by hearing protection

Role of Health Surveillance

This is appropriate for workers exposed to high noise levels, in the form of audiometry - a medical test that quantifies the sensitivity of a person's hearing. It normally involves the worker sitting in a soundproof booth while wearing headphones, listening for a series of beeps and pressing a button when they are heard. This maps the hearing capabilities across the range of frequencies.

Audiometry allows:

- Recognition of existing hearing loss (before starting employment).
- Further damage or hearing loss during employment to be identified.
- The removal or exclusion of workers from high noise areas (to protect from further loss).
- An evaluation of the effectiveness of noise controls.

Audiometry should be conducted by a trained, competent person. Good records must be kept and employees informed of the results of their checks. If any damage is identified, then the affected worker should be examined by a doctor.

STUDY QUESTIONS

1. What does a daily personal exposure of 85 dB(A) mean?
2. What are the general limitations of ear defenders and ear plugs?
3. Identify three benefits that audiometry allows.

(Suggested Answers are at the end.)

13.2 Vibration

Vibration

IN THIS SECTION...

- Exposure of the hands to excessive vibration can lead to hand-arm vibration syndrome. There are also health effects from whole-body vibration.
- Exposure standards exist for both hand-arm vibration and whole-body vibration.
- Exposure to vibration can be controlled by: mechanisation, low-vibration emission tools, selecting suitable equipment; maintenance programmes; and limiting exposure (including use of PPE).
- Health surveillance can protect workers from the effects of vibration and help to ensure controls are adequate.

The Effects on the Body of Exposure to Vibration

Vibration is similar in many respects to noise, both in terms of physical characteristics and preventive measures. The health effects associated with vibration exposure fall into two main categories.

Hand-Arm Vibration Syndrome (HAVS)

Regular and frequent use of vibrating tools and equipment is found in a wide range of construction activities, e.g. building and maintenance of roads and railways; concrete products; construction of mines and quarries; and house building. Any vibrating tool or process which causes tingling or numbness after five to ten minutes of continuous use is suspect. Exposure to such vibration can lead to Hand-Arm Vibration Syndrome (HAVS), the symptoms of which include:

- **Vibration White Finger (VWF)** - where the blood supply to the fingers shuts down and the fingers turn white. This is made worse by cold or wet conditions. The blood supply will return after some time, leaving the fingers red and painful. This is the most common form of HAVS.
- **Nerve damage** - the nerves to the fingers stop working properly, resulting in a loss of pressure, heat and pain sensitivity.
- **Muscle weakening** - grip strength and manual dexterity reduce.
- **Joint damage** - abnormal bone growth at the finger joints can occur.

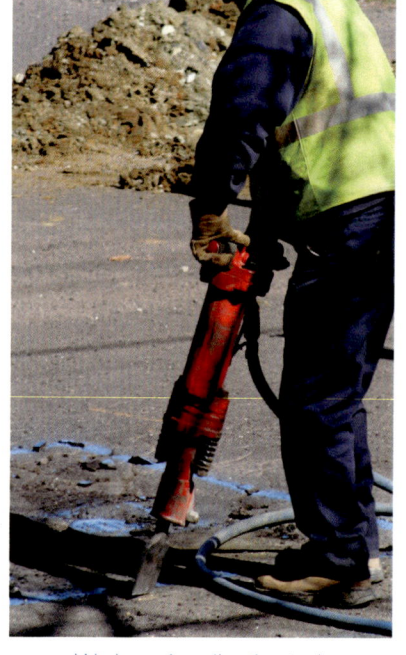

Worker using vibrating tool

HAVS usually takes many years (five to ten) of exposure to vibrating equipment to manifest itself, but once established it is not curable. Any further exposure to vibration will do further damage. The only effective solution is to stop an affected person from using vibrating tools.

HAVS is a notifiable disease under the **Reporting of Injuries, Diseases and Dangerous Occurrences Regulations 2013 (RIDDOR)**.

Vibration | 13.2

TOPIC FOCUS

Vibration Risk Activities

Work equipment that can expose workers to vibration risks includes:

- Chainsaws and circular saws.
- Needle guns and scrabbling machines.
- Road/rock drills, concrete breakers, hammer drills.
- Hand-held grinders, pedestal grinders and hand-held sanders.
- Nut runners.
- Power hammers and chisels, riveting hammers and bolsters.

Operators are particularly at risk if using:

- Hammer action equipment for more than half an hour each day.
- Rotary or other action equipment for more than two hours each day.

Whole-Body Vibration (WBV)

Not yet fully understood, Whole-Body Vibration (WBV) can result from using various construction vehicles or compactors, and can cause other injuries to the knees, hips and back. The damage is caused by vibration from the vehicle or machine passing through the seat into the driver's body through the buttocks. Additionally, standing on the platform of a machine will result in vibration passing through the worker's feet.

Effects will be worse if travelling too fast, through rough terrain, on badly paved surfaces, or if the vehicle itself has poor suspension. The greater the duration and level of vibration, the worse the effect may be. Contributory factors which can cause or increase the effects are poor driving posture, poor design of controls, or lack of visibility making twisting and/or turning necessary when driving.

Construction vehicle drivers or those operating large static compactors, hammering or punching machines and mobile crushers are most at risk from WBV. Young workers may also be susceptible in this respect.

The most significant health effect is back pain as a result of damage to the soft tissues of the spine (such as the intervertebral discs) though other effects have been reported (such as vertigo).

When Exposure Should be Assessed

The **Control of Vibration at Work Regulations 2005 (CVAWR)** provide the framework for the regulation of exposure of workers to vibration, in order to protect their health.

The employer needs to ascertain (by risk assessment) whether employees are liable to be exposed to vibration levels at or above an exposure action value or above an exposure limit value (see below). These limits are set on the basis that the amount of damage done is dependent on the vibration energy absorbed by the body. This is determined by the:

- Vibration magnitude, measured in m/s^2.
- Duration of the exposure, measured in hours and minutes.

These two factors determine the 'dose' of vibration absorbed (the same principle as applied to noise).

13.2 Vibration

> **TOPIC FOCUS**
>
> A **vibration risk assessment** needs to consider:
>
> - Equipment likely to cause vibration and places of use.
> - The employees, and the magnitude, type and duration of exposure.
> - The effects of vibration on employees who may be at risk.
> - Any effects of vibration on the workplace and equipment.
> - Manufacturers' information and vibration data.
> - Availability of replacement equipment with reduced vibration.
> - Any further exposure at the workplace to whole-body vibration beyond normal working hours.
> - Exposure in rest facilities supervised by the employer.
> - Specific working conditions, e.g. low temperatures.
> - Appropriate information obtained from health surveillance including, where possible, published information.
>
> The assessment should be recorded and reviewed.

Risk assessment should enable the employer to decide whether employees' exposure is likely to be above the **Exposure Action Value (EAV)** or **Exposure Limit Value (ELV)** (see below) and to identify work activities in need of control.

The assessment must be completed by a competent person who must:

- Have had specific training.
- Know:
  - The work processes being used.
  - How to collect and understand relevant information.
- Be able to develop a plan of action based on the findings.
- Ensure the plan is introduced and is effective.

Work activities involving vibrating equipment should be grouped according to the risk they represent: high, medium or low. Action should be planned to control risks for the employees at greatest risk first.

Rough groupings should be based on:

High Risk	Those operating: • Hammer action tools for more than 1 hour per day. • Rotary and other action tools for more than about 2 hours per day.
Medium Risk	Those who regularly operate: • Hammer action tools for more than 15 minutes per day. • Some rotary and other action tools for more than about 1 hour per day.

Comparison of Measurements to Exposure Limits Established by Recognised Standards

Exposure Action Value and Exposure Limit Value

CVAWR require employers to take specific action when the daily vibration exposure reaches a certain **action** or **limit** value.

The daily **Exposure Action Value (EAV)** is:

- For hand-arm vibration 2.5 m/s² A(8).
- For whole-body vibration 0.5 m/s² A(8).

If these are reached, the employer must:

- Carry out a vibration risk assessment.
- Reduce vibration exposure to the lowest level reasonably practicable.
- Provide information, instruction and training for employees.
- Carry out health surveillance where the assessment indicates a risk to health due to the EAV being exceeded.

The daily **Exposure Limit Value (ELV)** (which must not be exceeded) is:

- For hand-arm vibration 5.0 m/s² A(8).
- For whole-body vibration 1.15 m/s² A(8).

If these are reached, the employer must:

- Carry out a vibration risk assessment.
- Immediately reduce exposure below the ELV.

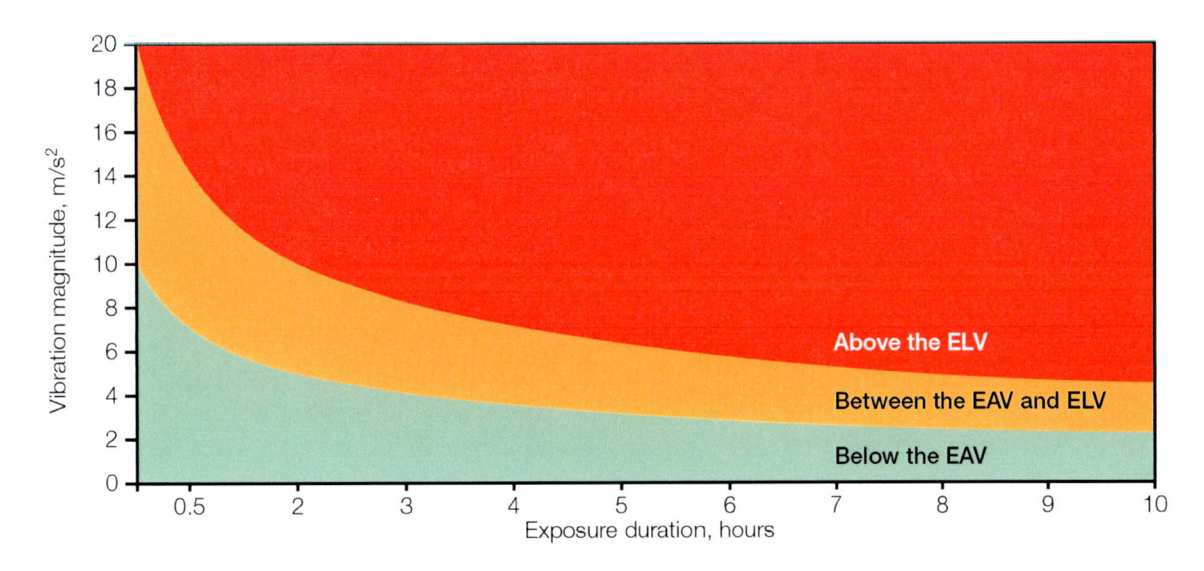

How vibration level and duration affect exposure
(Source: INDG175(rev2) Control the risks from hand-arm vibration, HSE, 2008)

13.2 Vibration

Basic Vibration Control Measures

There are a number of preventive and precautionary measures which can be taken:

- **Mechanisation**
    - Mechanise the activity - use a concrete breaker mounted on an excavator arm rather than hand-operated.
- **Low-Vibration Emission Tools**

    Anti-Vibration Technology (AVT) is an emerging science which is being applied to a wide range of modern construction tools. Other methods for reducing vibration are to:
    - Change the tool or equipment for one with less vibration generation characteristics.
    - Use tools that create less vibration, e.g. a diamond-tipped masonry cutter instead of a tungsten hammer drill.
    - Support the tools (e.g. tensioners or balancers), allowing the operator to reduce grip and feed force.
    - Add anti-vibration mounts to isolate the operator from the vibration source.
- **Selection of Suitable Equipment**

    Technological advances have improved designs of hand-held machines so that they have lower vibration levels. Buying newer, low-vibration tools will reduce vibration. However, simply buying new tools may not be the answer because the vibration emissions may still present a risk.

    The tool's efficiency is important. It may be too powerful for the job causing unnecessarily high-vibration exposure. It may take a long time to complete a job - compared with using a more efficient tool with a greater vibration emission - and so increase exposure to higher vibration levels.
- **Maintenance Programmes**
    - Keep moving parts properly adjusted and lubricated.
    - Keep cutting tools sharp.
    - Replace vibration mounts before they wear too badly.
    - Ensure rotating parts are checked for balance.
    - Keep all equipment clean - especially look for corrosion.
- **Limiting Exposure**
    - Calculate how long a tool/job takes to reach an action level or limit.
    - Operate tools within these known action levels or limits.
    - Avoid gripping too tightly or forcing tools.
    - Use job rotation techniques - share between several workers.
    - Ensure adequate rest breaks during the work.
- **Suitable PPE**
    - PPE (e.g. gloves) may not actually protect against vibration.
    - Gloves will protect from cold and wet - a major contributing factor.

Further measures to minimise whole-body vibration:

- Maintain suspension components adequately.
- Ensure the driver's seat is in good repair and gives adequate support.
- Plan site routes to use the smoothest terrain.
- Fit a suspension seat, correctly adjusted for the driver's weight.
- Make sure correct tyre pressures are used.
- Adjust speed where the terrain is rough.

Role of Health Surveillance

Regulation 7 of **CVAWR** requires that health surveillance should be conducted where appropriate, e.g. in cases where the risk assessment shows a risk of developing vibration-related conditions, or employee exposure reaching action levels. Records of this health surveillance should be kept.

Where an identifiable disease related to vibration exposure is discovered, monitoring is required to minimise the health effects and maintain adequate control. The employer needs to take the following steps:

- Ensure that a qualified person informs the employee accordingly and provides them with information and advice.
- Ensure that the employee is informed of any significant findings.
- Review the risk assessment.
- Review any measures taken to control risk from vibration.
- Consider assigning the employee to alternative work where there is no risk from further exposure to vibration.
- Provide for a review of the health of any other employee who has been similarly exposed.

STUDY QUESTIONS

4. State the symptoms of hand-arm vibration syndrome.
5. What control measures can be taken with regard to tools and equipment, to reduce the risk of vibration injuries?
6. When do the **Control of Vibration at Work Regulations 2005** require that health surveillance be conducted?

(Suggested Answers are at the end.)

13.3 Radiation

Radiation

IN THIS SECTION...

- Radiation comes in two forms - ionising radiation and non-ionising radiation.
- Ionising radiation includes alpha particles, beta particles, X-rays, gamma rays, radon and neutrons. Exposure to these can cause acute sickness and chronic effects.
- Non-ionising radiation includes ultraviolet (UV), infrared (IR), visible light, microwaves and radio waves. UV, visible light and IR can cause eye and skin damage; microwaves and radio waves can cause internal heating.
- Artificial optical radiation includes sources of non-ionising radiation, and an employer must consider the causes and effects of these in risk assessment.
- Exposure to ionising radiation can be controlled by properly managing time and distance and by providing shielding. Dose limits apply.
- Clothing and PPE can help to control exposure to non-ionising radiation, as can engineering controls such as restricted access.
- Radiation protection strategies are required where radioactive sources are used in the workplace.

Differences Between Types of Radiation and their Health Effects

Radiation is energy that is emitted from a radioactive source. The energy emitted is capable of causing considerable harm, depending on its form and the length of exposure. The higher the frequency, the more penetrative its properties.

There are two forms of radiation - **ionising** and **non-ionising**.

DEFINITIONS

IONISING RADIATION

Radiation that causes ionisation in the material that absorbs it.

NON-IONISING RADIATION

Radiation that does not cause ionisation in the material that absorbs it.

Ionising and non-ionising radiation can both be divided into several main types:

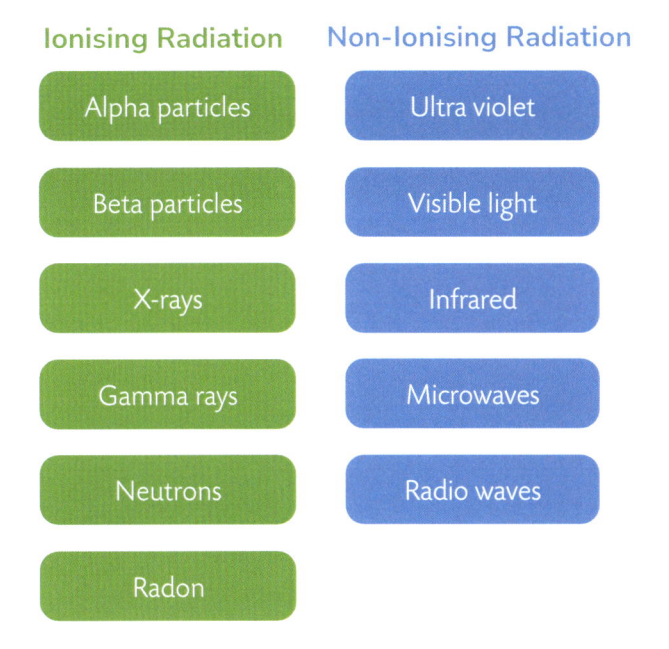

Ionising Radiation	Non-Ionising Radiation
Alpha particles	Ultra violet
Beta particles	Visible light
X-rays	Infrared
Gamma rays	Microwaves
Neutrons	Radio waves
Radon	

Ionising Radiation

The main types are:

- **Alpha Particles**
    - Sub-atomic particles.
    - Low penetration, stopped by thin barriers such as paper and skin.
    - Low hazard outside the body; higher risk if swallowed or inhaled.
- **Beta Particles**
    - Sub-atomic particles.
    - More penetration and can get through skin to living tissue.
    - Hazardous outside the body.
- **X-rays**
    - High-energy electromagnetic light emitted from radioactive substances and generators (X-ray machines).
    - High penetrating power and can shine right through the human body (not through dense bone tissue).
    - Very hazardous.
- **Gamma Rays**
    - Very high electromagnetic energy (light) emitted by some radioactive substances.
    - Very high penetration, can go right through the human body (even through bones).
    - Can penetrate solid objects such as steel and concrete to a degree.
    - Very hazardous.
- **Neutrons**
    - Sub-atomic particles emitted by some radioactive substances.
    - Very high penetrating power, can penetrate through the body.
    - Very hazardous.

13.3 Radiation

- **Radon**
  - More properly known as Radon 222.
  - A naturally occurring colourless and odourless radioactive gas.
  - Decay products of radon are themselves radioactive (alpha particles).
  - Very hazardous.

Health Effects of Ionising Radiation

Acute effects of exposure to high doses include:

- Sickness and diarrhoea.
- Hair loss.
- Anaemia, due to red blood cell damage.
- Reduced immune system due to white blood cell damage.

All of the cells of the body are affected by the radiation, but some more than others. A large enough dose can kill in hours or days.

Chronic effects of exposure include:

- Cancer.
- Genetic mutations.
- Birth defects.

Chronic effects can arise following exposure to high and low doses of radiation. There is no safe level of exposure below which there will be no chronic effects - instead, there is a clear relationship between dose and risk (i.e. the higher the dose, the higher the risk).

Effects of radon - most radon gas that is inhaled is immediately exhaled and will cause little harm. However, decay products of radon act more like solid particles and are themselves radioactive. These solid decay products attach to atmospheric dusts and water droplets, which, if inhaled, become lodged in the airways and the lungs. Some emit alpha radioactive particles, which cause significant damage to sensitive lung cells.

Radon is the second largest cause of lung cancer in the UK (after smoking) accounting for 2,000 fatalities a year.

Non-Ionising Radiation

This form consists of lower-energy electromagnetic waves, whose energy decreases with increasing wavelength. There is a spectrum of types of non-ionising radiation based on the wavelength of the energy transmitted. This spectrum, together with the effects on the body, is as follows:

- **Ultraviolet (UV)**

  High-frequency electromagnetic radiation (light) emitted from white-hot materials (e.g. the arc produced in arc welding).

- **Infrared (IR)**

  Lower-frequency electromagnetic radiation (light) emitted by red-hot material (e.g. molten metal poured into castings).

- **Visible Light**

  Electromagnetic radiation between the UV and IR frequencies and visible to the human eye.

- **Radio Waves**

  Lower-frequency electromagnetic radiation emitted by an antenna.

- Microwaves

    Lower-frequency electromagnetic radiation emitted by a microwave generator (can be categorised as a subset of radio waves).

Lasers are sources of non-ionising radiation, and can operate at UV, visible light and IR frequencies. Operating at UV and IR, the rays are not visible to the human eye. Laser light is very coherent - the light waves are all aligned with one another and do not spread out over distance. They can carry power over long distances.

Artificial Optical Radiation

Artificial optical radiation is not a form of radiation in itself, but includes many of the non-ionising sources we have seen above (ultraviolet, infrared, lasers, etc.). The **Control of Artificial Optical Radiation at Work Regulations (AOR) 2010** require employers to give more consideration to controlling exposure to such artificial sources, to ensure they cause no harm.

Sources of **artificial optical radiation** in the workplace will include:

- Ceiling mounted lights (bulbs or tubes).
- All task lighting such as desk lamps.
- Tungsten-halogen lamps.
- High-pressure mercury floodlights.
- Photocopiers.
- Computer, phone and tablet screens.
- LED remote control devices and LED lamp systems.
- Flash lamps on cameras.
- Gas-fired overhead heaters.
- Lights, indicator lamps and headlights on vehicles.
- Certain classes of lasers.
- Desktop digital projectors.
- Welding and burning (arc and oxy-fuel).
- UV insect traps.

Headlights on vehicles are a source of artificial optical radiation which can often be present in construction activities

The sources identified above are generally considered to be 'safe' - in that, if used correctly, and not positioned in very close proximity to the eyes or skin, they will not cause harm.

Health Effects of Non-Ionising Radiation

These depend on and differ with the types of radiation:

- **UV**
    - Redness and skin burns (e.g. sunburn).
    - Pain and inflammation to the surface of the eye, leading to temporary blindness (often called 'arc-eye' or 'snow-blindness').
    - Increased risk of skin cancer.
    - Premature aging of the skin.

- **Visible Light**

    Can cause temporary blindness if intense (disability glare) and permanent eye damage if very intense (e.g. high-powered laser) and burns to exposed skin tissue.

13.3 Radiation

- IR
  - Redness and burns to the skin.
  - Development of eye cataracts over time.
  - Note: sunlight includes UV, visible light and IR - see above for effects.
- Microwaves
  - Absorbed into the body and cause internal heating.
  - High doses cause internal organ damage and could be fatal.
- Radio Waves

  Are absorbed and cause internal heating as microwaves.

Lasers are classified according to intrinsic safety and power output. Class 1 lasers present little risk, but Class 4 lasers can cause instant skin burns and eye damage.

Artificial optical radiation effects:

- Burns or reddening of the skin.
- Burns or reddening of the surface of the eye (photokeratitis).
- Burns to the retina of the eye.
- Blue-light damage to the eye (photoretinitis).
- Damage to the lens of the eye that causes early cataracts.

Typical Occupational Sources of Radiation

Ionising Radiation

This is used to provide energy, take X-rays, detect contaminants (radioactive (sealed) sources are used as portable nuclear density/moisture gauges to measure density and detect moisture in construction materials). Sources include the following:

Alpha particles	Smoke detectors and science labs.
Beta particles	Science labs and thickness gauges.
X-rays	Medical radiography, baggage security scanners, non-destructive testing of equipment and machinery.
Gamma rays	Industrial radiography - employing highly radioactive materials such as cobalt-60 is used commonly on construction sites. The process of gamma radiography - a type of non-destructive testing (NDT) - is used to validate the integrity of poured concrete.
Neutrons	Nuclear power stations.

Radon (Radon 222)

Radon is a naturally occurring radioactive gas originating from uranium, occurring naturally in rocks and soils - radon levels are much higher in certain parts of the UK. The highest levels are found in underground spaces such as basements, caves, mines, utility industry service ducts and in some areas in ground floor buildings, as they are usually at a higher pressure than the surrounding atmosphere. It usually gets into buildings through gaps and cracks in the floor.

All workplaces can be affected in radon-affected areas.

Non-Ionising Radiation

Sources of non-ionising radiation (including artificial optical radiation) include:

UV	Sunlight; arc-welding and oxy-fuel welding/burning; curing of paint in manufacturing and vehicle painting processes; curing of inks in printing.
Visible light	Laser levelling devices; laser pointer.
IR	Red-hot steel in a rolling mill; glass manufacture; ceramics (clay ware) manufacture.
Microwaves	Food processing (ovens); telecommunications equipment (mobile phone masts).
Radio waves	Radio, TV or radar transmitters.

Basic Ways of Controlling Exposure to Radiation

The **Ionising Radiations Regulations 2017 (IRR)** provide the framework for controls.

Protection from **ionising radiation** is based on strict adherence to forms of control which limit exposure to the absolute minimum by using time, distance and shielding.

Protection from the effects of **non-ionising radiation** is principally in the form of shields and PPE, although where ultraviolet sources are powerful enough to constitute a hazard, administrative controls may also be necessary.

Controlling Exposure to Ionising Radiation

- **Time** - minimise the duration of exposure. The dose is proportional to time - halve the time, halve the dose.
- **Distance** - the dose will get lower the further away from the source you get.
    - Alpha and beta particles only travel a short distance through air, so short separation distances are often effective.
    - X-rays and gamma rays travel much further, but obey the inverse square law - if the distance from the source to the person is doubled, the radiation dose goes down to a quarter (not a half).
- **Shielding** - relatively thin shields can be used with alpha and beta particles, but X-rays and gamma rays need much thicker, denser shields, such as lead.

Control of Radioactive Substances

Radioactive substances used as a source of ionising radiation should be in the form of a sealed source whenever reasonably practicable. This should be designed, constructed and maintained to prevent leakage. Records of the quantity and location of radioactive substances must be kept.

Controlling Exposure to Radon

A survey should be undertaken to determine if radon levels are acceptable or require action.

Where radon levels are at or above the action level of 400Bq/m^3 (becquerels per cubic metre), then employee exposure must be reduced. This is often best done by appointing a Radiation Protection Adviser (RPA) to carry out a risk assessment (see later).

Engineering solutions to high radon levels can often be applied, such as:

- Fitting positive pressure air fans to prevent the radon gas from seeping from the ground up into the workplace.
- Installing radon sumps and extraction systems to draw radon out of the ground at low-level before it can seep into buildings.

13.3 Radiation

Controlling Exposure to Non-Ionising Radiation

Control of exposure to non-ionising radiation is generally through the use of barriers and PPE. However, prolonged exposure to ultraviolet radiation may need warning signs, access restrictions and limited exposure times. All require information, instruction and training.

Engineering controls:

- Restricted access to controlled and supervised areas by interlocked doors.
- Where UV or visible radiation may be generated, reflective surfaces should be avoided and surfaces painted in a dark, matt colour.
- Segregation and containment of unsealed radioactive material.
- Shielding is the best method of protection.
- There should be no unnecessary metal objects near any radiating Radio Frequency (RF) device, as localised high field strengths may result around such items. Care should therefore be taken to remove rings, watches or bracelets when working close to radiating sources.

Radiation detector

PPE:

- Gloves and overalls prevent exposure from low-energy beta emitters and prevent skin contact.
- High-density materials are used to provide shielded body protection for persons at risk from penetrating radiation, e.g. radiographers.
- Eye protection, possibly using high-density lenses, may be used to protect the eyes if the head is at risk from exposure to beams of radiation.
- RPE may be needed as an additional precaution to prevent inhalation of radioactive contamination.

Controlling Exposure to Artificial Optical Radiation

A simple hierarchy of control measures should include:

- Use an alternative, safer light source that will achieve the same result.
- Use filters, screens, remote viewing, curtains, safety interlocks, clamping of workpieces, dedicated rooms, remote controls and time delays.
- Train workers in best practice and provide appropriate information.
- Organise work to reduce exposure to workers.
- Restrict access to hazardous areas.
- Issue PPE (as above).
- Display relevant safety signs.

Should harmful over-exposure occur, quick referral to a medical practitioner or occupational health provider will be required. Remember to include AOR in your risk assessment of exposure to non-ionising radiation.

Basic Radiation Protection Strategies

Radiation Protection Adviser

Where controlled areas have been designated, the employer must appoint a qualified Radiation Protection Adviser (RPA). Usually from external organisations, RPAs must have particular experience of the type of work the employer undertakes and be able to provide advice and guidance on the following matters:

- Compliance with current legislation.
- Local rules and systems of work.

- Personnel monitoring, dosimetry and record-keeping.
- Room design, layout and shielding.
- Siting of equipment emitting ionising radiation.
- Siting and transport of radioactive materials.
- Leakage testing of sealed sources.
- Investigation of incidents, including spillage or loss.

Local Rules

Employers must provide local rules to describe the safe systems of working with ionising radiation. These must be prominently displayed and brought to the attention of all relevant employees.

Radiation Protection Supervisor

One or more Radiation Protection Supervisors (RPSs) must be appointed (usually internally) to be responsible for enforcing the local rules. The duties of the RPS include:

- Record-keeping.
- Registration of workers.
- Radiation monitoring.
- Implementation of local rules.

Controlled and Supervised Areas

Areas where there is a radiation hazard must be designated by the RPA and access restricted to classified workers to reduce the radiation dose.

At the entrance to each area, a sign must be posted indicating the:

- Designation of the area.
- Nature of the source and any restriction of activities.
- Names of authorised workers.
- Supervisors of work in the area.

Within the area, all sources of radiation must be clearly marked with the radiation hazard symbol.

Examples of common radiation hazard-warning signs

Two categories of area for radiation work are determined by the likely radiation dose of those working in the area:

- **Controlled Area**

    Exposures exceeding three tenths of a dose limit may be received. These areas will include where radioactive materials are stored and dispensed.

- **Supervised Area**

    Where a worker could receive between one tenth and three tenths of a dose limit. Most areas in which work with radiation is carried out will normally be supervised areas, including X-ray rooms.

13.3 Radiation

A safe system of work will include rules for handling radioactive source material, action in the event of accidents or other incidents, and procedures on leaving a controlled or supervised area. A formal permit-to-work system may be required to restrict time spent in the radiation area. Areas designated as controlled or supervised areas must have washing and changing facilities.

The Role of Monitoring and Health Surveillance

The International Commission on Radiological Protection (ICRP) has set the following dose limits on exposure to ionising radiation:

- The general public shall not be exposed to more than 1 mSv (millisievert) per year.
- Occupational exposure shall not exceed 20 mSv per year.

These limits exclude exposure due to background and medical radiation.

To give you an idea of what these figures mean, the total natural radiation to which people are likely to be exposed is about 1.5 mSv per year; a chest X-ray involves an exposure of 0.04 mSv.

Three forms of monitoring are used:

- **Personal monitoring** - personal dosimeters used for those in controlled and supervised areas. These may measure whole body dose or partial body dose (i.e. the fingers).
- **Medical examination** - routine examinations conducted before employment and every 12 months, with an immediate special examination after an over-exposure.
- **Area monitoring** - levels of radiation in controlled and supervised areas must be regularly assessed and monitoring equipment properly maintained, examined and tested by a competent person every 14 months.

Records of all forms of monitoring must be kept.

Artificial Optical Radiation

Where any employees may be over-exposed to sources of artificial optical radiation (such as IR, UV, etc.), an employer should provide medical examination and consider whether follow-up health surveillance is appropriate.

STUDY QUESTIONS

7. What type of non-ionising radiation is given off by the following pieces of equipment?
    (a) Radio transmitter.
    (b) Hot plate on a stove.
    (c) Arc welder in operation.
    (d) Laser measuring device.
8. What are the acute health effects of high doses of ionising radiation?
9. What are the health effects of artificial optical radiation?
10. What is radon, and where is it likely to be found?
11. Identify four areas for which a Radiation Protection Adviser (RPA) is responsible, and outline when an employer must appoint one.

(Suggested Answers are at the end.)

Mental Ill Health

IN THIS SECTION...

- Common symptoms of mental ill health are: depression, anxiety/panic attacks and Post-Traumatic Stress Disorder (PTSD).
- Mental ill health can be caused by: unreasonable demands, lack of control, lack of support, poor working relationships, an ill-defined role and change.
- To minimise the risk of serious ill health caused by stress, the employer should establish a management framework for: demands, control, support, relationships, role and change.

The Frequency and Extent of Mental Ill Health in the Construction Industry

According to HSE figures for 2020/2021, there were an estimated 20,000 work-related cases of stress, depression or anxiety (new or long-standing), 27% of all ill health in the construction industry.

Suicide is the biggest killer of men under the age of 45. However, male site workers are **three times** more likely to commit suicide than the average male in the UK.

Suicide kills more construction workers than falls.

Recognising Common Symptoms

Stress is not a disease, but an adverse natural reaction to pressure, and pressure is an inherent part of work, e.g. a deadline to be met or an order to get out on time. Some even say that there are 'good' pressures, that give us strength when needed (the 'fight or flight' response); but never 'good stress'.

If demands are placed on workers which they feel they cannot cope with, they will experience stress, which in turn affects morale and performance. People react differently and some will see certain events as being stressful, while others may not.

Depression

Chronic stressful life situations can increase the risk of developing depression. The symptoms of depression can be complex and vary widely between people.

The **psychological** symptoms of depression include:

- continuous low mood or sadness;
- feeling hopeless and helpless;
- having low self-esteem;
- feeling tearful;
- feeling guilt-ridden;
- feeling irritable and intolerant of others;
- having no motivation or interest in things;
- finding it difficult to make decisions;
- not getting any enjoyment out of life;
- feeling anxious or worried.

13.4 | Mental Ill Health

The **physical** symptoms of depression include:

- moving or speaking more slowly than usual;
- changes in appetite or weight (usually decreased, but sometimes increased);
- constipation;
- unexplained aches and pains;
- lack of energy;
- low sex drive (loss of libido);
- changes to your menstrual cycle;
- disturbed sleep.

The **social** symptoms of depression include:

- avoiding contact with friends and taking part in fewer social activities;
- neglecting your hobbies and interests;
- having difficulties in your home, work or family life.

Anxiety/Panic Attacks

Panic attacks are generally more intense than anxiety attacks. They also come on out of the blue, while anxiety attacks are often associated with a trigger.

While panic attacks come on suddenly, symptoms of anxiety follow a period of excessive worry.

Symptoms may become more pronounced over a few minutes or hours. They are typically less intense than those of panic attacks.

Anxiety attack symptoms include:

- being easily startled;
- chest pain;
- dizziness;
- dry mouth;
- fatigue;
- fear;
- irritability;
- loss of concentration;
- muscle pain;
- numbness or tingling in the extremities;
- a rapid heart rate;
- restlessness;
- shortness of breath;
- sleep disturbances;
- the feeling of being choked or smothered;
- worry and distress.

Anxiety symptoms often last longer than the symptoms of a panic attack. Panic attacks come on suddenly, without an obvious trigger. Symptoms include:

- a racing or pounding heartbeat;
- chest pain;
- dizziness or lightheadedness;
- hot flashes or chills;
- nausea;
- numbness or tingling in the extremities;
- shaking;
- shortness of breath;
- stomach pain;
- sweating;
- the feeling of being choked or smothered.

Post-Traumatic Stress Disorder (PTSD)

Post-Traumatic Stress Disorder (PTSD) is a type of anxiety disorder which you may develop after being involved in, or witnessing, traumatic events. The specific symptoms of PTSD can vary widely between individuals, but generally fall into these three categories:

- Re-experiencing - this is when a person involuntarily and vividly relives the traumatic event in the form of flashbacks and nightmares.
- Avoidance and emotional numbing - this usually means avoiding certain people or places that remind you of the trauma, or avoiding talking to anyone about your experience.
- Hyperarousal (feeling 'on edge') - someone with PTSD may be very anxious and find it difficult to relax. They may be constantly aware of threats and easily startled. It often leads to irritability and angry outbursts.

Causes of and Controls for Mental Ill Health

Causes

The areas to consider for both causes of stress and prevention are the same: demands, control, support, relationships, role and change.

- **Demands**

    Excessive demands of the job in terms of:

    - Workload (too much or too little).
    - Speed of work and deadlines.
    - Excessively long working hours.
    - Work patterns (changing shift patterns).

    Also consider the nature of the job:

    - Some are inherently difficult (e.g. air traffic control).
    - Some expose workers to highly emotional situations (e.g. social work).

Excessive workload

13.4 Mental Ill Health

- **Control**

    Lack of control over:

    - How work is done.
    - The priorities involved.
    - Simple issues such as the environment (e.g. light levels, temperature, background noise).

- **Support**

    Lack of support in terms of:

    - Information, instruction and training to do the work.
    - Having no-one to turn to for help when pressure increases.

- **Relationships**

    Poor workplace relationships, and, in particular, bullying and harassment (whether by managers, colleagues, subordinates, customers, suppliers or members of the public).

- **Role**

    - Lack of clarity about an individual's role, what the responsibilities and requirements of the job are (role ambiguity).
    - Subject to conflicting demands (role conflict) and how they fit into the larger organisational structure.

- **Change**

    The threat of change and the change process itself, be it a change that affects only one person (e.g. demotion, re-assignment) or the whole organisation (e.g. redundancies, take-over), can create huge anxiety and insecurity.

There are many things in a person's life outside of work that can cause stress just as badly as work can, and these factors (psycho-social) should also be considered. Even though they are not directly caused by work, they are brought into the workplace, and can therefore affect both the person experiencing the difficulties, and others.

MORE...

The Management Standards for Work Related Stress can be found at:

www.hse.gov.uk/stress/standards/

Control Measures

Case law emphasises the particular need to minimise stress, especially for an employee with a record of stress-related illness. Strategies to tackle work-related stress must be based on a risk assessment and the principles set out in the **Management of Health and Safety at Work Regulations 1999**:

- Avoiding the risks.
- Combating the risks at source.
- Adapting the work to the individual.
- Developing collective measures.

The basic management framework should be based around the same issues listed by the HSE as causes of stress:

- **Demands**

    - The demands (in terms of workload, speed of work and deadlines) should be reasonable and, where possible, set in consultation with workers.
    - Working hours and shift patterns should be carefully selected and flexible hours allowed where possible.
    - Workers should be selected on their competence, skills and ability to cope with difficult or demanding work.

- Control

    Employees should be encouraged to have more say in how their work is carried out, e.g. in planning their work, making decisions about how it is completed and how problems will be tackled.

- Support
    - Feedback to employees will improve performance and maintain motivation.
        - All feedback should be positive, with the aim of bringing about improvement, even if this is challenging.
        - Feedback should focus on behaviour, not on personality.
    - Workers should have adequate training, information and instruction.
- Relationships
    - Clear standards of conduct should be communicated to employees, with managers leading by example.
    - The organisation should have policies in place to tackle misconduct, harassment and bullying.
- Role
    - An employee's role in the organisation should be defined by means of an up-to-date job description and clear work objectives and reporting responsibilities.
    - If employees are uncertain about their job or the nature of the task to be undertaken, they should be encouraged to ask at an early stage.
- Change
    - If change has to take place, employees should be consulted about what the organisation wants to achieve and given the opportunity to comment, ask questions and get involved.
    - Employees should be supported before, during and after the change.

Recognition That Most People with Mental Ill Health Can Continue to Work Effectively

Everyone has mental health, just as we all have physical health. During the course of a lifetime, everyone will face challenges to their mental wellbeing just as we face challenges to our physical health. Employees with mental ill health issues can still perform effectively but some adjustments to their working conditions and environment may need to be made. By law, employers must make 'reasonable adjustments' for workers with disabilities or long-term physical or mental conditions.

This might involve flexible working hours, support from colleagues, a place to rest, etc.

Organisations That Provide Support

A number of organisations offer mental health advice and support across the UK, including:

- Fit for Work.
- Mind.
- Mental Health Foundation.
- Rethink.

13.4 Mental Ill Health

TOPIC FOCUS

Reducing levels of occupational stress in construction work:

- Ensure policies and procedures are in place to cover harassment, discrimination, violence and bullying.
- Have managers and supervisors trained to recognise the symptoms of stress.
- Reduce levels of noise on site.
- Ensure adequate lighting in all work areas (especially at night).
- Ensure adequate welfare and rest facilities are available on site.
- Ensure high standards of housekeeping and maintenance of equipment and facilities.
- Discourage long working hours.
- Introduce job rotation and increase work variety as much as possible.
- Ensure competence and ensure people are properly matched to jobs.
- Maintain high levels of communication and involve workers in making decisions.

STUDY QUESTIONS

12. State the six main work-related causes of mental ill health and, for each, give one example of a preventive measure.
13. Identify how to reduce occupational stress in construction work.

(Suggested Answers are at the end.)

Violence at Work

IN THIS SECTION...

- Violence at work is any incident where a worker is abused, threatened or assaulted while working.
- Various factors influence the risk of work-related violence and many occupations are at risk. The chance of violent incidents is notably higher in certain roles - particularly those which deal with the public.
- The risk of violence at work can be controlled using a control hierarchy: avoid the risk through elimination/substitution, use engineering controls, introduce procedural measures and introduce individual measures.
- All violent incidents should be investigated to avoid similar occurrences in future.

Introduction to Violence at Work

DEFINITION

WORK-RELATED VIOLENCE

Any incident in which a person is abused, threatened or assaulted in circumstances relating to their work.

Findings from the 2019/20 Crime Survey for England and Wales (CSEW) show that:

- 307,000 adults experienced work-related violence.
- There were an estimated 688,000 incidents of work-related violence including threats and physical assault.
- Strangers were the offenders in 60% of cases of workplace violence.
- Out of the other 40% of incidents where the offender was known, they were most likely to be a client or a member of the public known through work.
- In 2020/2021, the **Reporting of Injuries, Diseases and Dangerous Occurrences Regulations 2013 (RIDDOR)** reported 4,286 injuries to employees, where the 'kind of accident' was 'physical assault/act of violence' in Great Britain (England, Wales and Scotland). This represents 8% of all reported workplace injuries. Of this figure, there were two deaths.

Certain construction occupations and types of work are associated with an increased risk of violence. The following factors relate to those occupations.

Types of Violence at Work

Physical

Physical attacks on workers in the construction industry are uncommon but when they do happen they can leave the victim with physical and mental scars. Situations that increase the risk factor are:

- **Handling high-value goods**, e.g. cash, construction materials or other valuable items - here the risk of violence is associated with robbery, e.g. from a site office handling cash transactions or holding petty cash.
- **Refusing** customers or clients, e.g. a loan arranged through a site office; a benefit or some other product (such as a free landscaped garden, patio or fire-alarm system included in the price for certain clients); access to another person; or compensation for a grievance.
- **Censuring** clients or customers, e.g. excluding them from site for breaches of rules; an inspector serving a prohibition notice on a site agent, contractor, etc.; or asking someone to leave the premises (e.g. a subcontractor due to poor workmanship).
- **Contact with customers/clients** who are under stress (perhaps as a result of frustrations or delays in obtaining the necessary contact), under the influence of alcohol or drugs, or with a history of violence. Any of these factors may serve to aggravate the above risk situations, or may be the cause of the risk itself.

13.5 Violence at Work

- **Dealing with employees** who may be subject to disciplinary measures, or may be under the influence of alcohol or drugs.

Note that reference is made to customers and clients - i.e. persons from outside the organisation. This forms the greatest risk, but there are also situations where workers may be at risk from other workers in their own organisation.

Violence amongst employees on construction sites is not common, however there are a number of factors that can increase this risk, e.g. money, illness or stress due to personal problems.

Other risk factors in respect of construction situations include:

- Staff handling situations alone, e.g. a night-watchman on a construction site.
- The time of day, where late evening/night workers are more vulnerable.
- The geographical area, e.g. work in urban areas is generally more risky and, within cities, particular areas have more social problems and higher crime rates, which may indicate a higher risk for workers in those areas.

A stressful situation can sometimes spill over into abuse, threat and assault

Psychological

The characteristics of emotional abuse are:

- It involves non-physical behaviour, which may range from delivering threats and insults to openly doling out public humiliation and intimidation.
- The abuse is intentional.
- The abuse is regular.

Verbal

Verbal abuse is defined as language that is intimidating, threatening or humiliating. Although some verbal abuse is quite obvious, it can also be more subtle. Examples of verbal abuse that can occur in the workplace include:

- Ridiculing, screaming and yelling at a colleague, for example, by calling them stupid or incompetent.
- Discussing a colleague in a gossiping manner with others, for example, by spreading rumours.
- Interrupting conversations with a colleague.
- Threatening a colleague.

Bullying

There are many ways to define bullying, with no single definition used across the board. The Advisory, Conciliation and Arbitration Service (ACAS) suggests the following: *"Offensive, intimidating, malicious or insulting behaviour, an abuse or misuse of power through means intended to undermine, humiliate, denigrate or injure the recipient."*

Bullies are often devious, operating out of sight of witnesses, and engaging methods which, when viewed in isolation, can seem fairly harmless. Bullying is almost always psychological, except in rare cases with male bullies. The target is often discriminated against because they are competent or popular.

Harassment

Definitions of harassment tend to refer to behaviour which is offensive and intrusive, with a sexual, racial or physical element.

ACAS defines harassment as: *"Unwanted conduct that violates people's dignity or creates an intimidating hostile, degrading, humiliating or offensive environment."*

Harassment usually has a strong physical component – e.g. contact, touch, intrusion into personal space, damage to possessions and sabotage of a target's work - and often takes place in public as a means of peer approval or image building.

An employee who is guilty of harassment will use individual difference as a means of victimising others. Their behaviour tends to be much more obvious, e.g. they may use offensive language, or harass the target in front of others. Harassment is, therefore, often easier to identify and confront.

Effective Management of Violence at Work

The HSE guide *INDG69 Violence at work - A guide for employers* recommends that employers adopt a four-stage approach.

- **Stage 1: Finding out if you have a problem:**
    - Ask the staff.
    - Keep detailed records.
    - Classify all incidents.
    - Try to predict what might happen.
- **Stage 2: Deciding what action to take**:
    - Decide who might be harmed, and how.
    - Evaluate the risk.
    - Record your findings.
    - Review and revise your assessment.
- **Stage 3: Take action**:
    - Develop a policy for dealing with violence and ensure that all employees are aware of it.
- **Stage 4: Check what you have done**:
    - Consult with employees to make sure the arrangements are working.

Interviews should take place to establish the nature of the problem

Other preventive measures, which will be different depending on the nature of the workplace and of the work, include the following:

- Elimination/Substitution

    Avoidance of the risk at source is the first strategy and this may be pursued through changing working practices:
    - **Minimisation of cash handling** - encouraging the use of cheques, credit and debit card transactions; and regular removal of any cash from site office, etc. to safe storage.
    - **Minimisation of customer/client frustration** - this will include measures to enable quick access to the point at which a person's problem or issue is properly dealt with, for example:
        - Providing sufficient staff to deal with numbers of customers/clients.
        - Site and site-office opening hours tailored to suit customers/clients.
        - Preliminary screening of people coming on site to ensure customers are routed to the correct contact person.
        - Improved reception and waiting facilities - including attention to decor and seating, etc.
        - Improved information - both at the point of contact and prior to it, e.g. the provision of clearly written instructions, information and explanations; easy access to contacting the right staff by telephone, etc.
    - **Refusing access to potentially violent customers and clients** - this enables the risk to be confined to security staff specially employed to perform such work.

13.5 Substance Abuse at Work

- **Engineering Controls**

    There are a range of physical security measures both to protect cash and valuables, and to protect staff themselves. Examples include:

    - Secure doors with coded entry locks to prevent unauthorised access.
    - Surveillance, video and alarm systems, including closed circuit video, etc.
    - Elimination of all unobserved and unused areas in premises and on site, particularly those from where there is no exit.
    - Improved lighting.
    - Removal or securing of loose objects which could serve as projectiles or weapons.
    - High or wide counters, or the use of security screens, to provide a physical separation between staff and customers/clients.

- **Procedural Measures**

    Systems of work may also be reviewed to ensure greater security of staff, e.g. accompanying staff where necessary and avoiding lone work (or having arrangements to keep in touch with lone workers).

- **Individual Measures**

    Measures at the personal level to protect staff include training and information, e.g. on how to recognise the early signs of aggression, how to deal with difficult customers/clients, etc. Staff may also be issued with personal protection in the form of personal alarms and communication devices (including mobile phones).

- **Investigation**

    Violent incidents must always be properly investigated and support offered to the victim. Investigations should determine whether procedures were adequate or more are required (e.g. information and changed working practices). It is important that the victim of any violent act is debriefed, has appropriate time off work/counselling, and legal help, and that other employees are given training/assistance to help them react to any such situation that arises.

- **Preventing Violence Among Employees**

    Good supervision allied with robust selection procedures for employees, contractors and subcontractors will help to alleviate possible problems. During this process, adequate information, instruction and training regarding disciplinary procedures and their consequences for all concerned are required. Effective consultation and commitment and the provision of information on the company policy and actions to take in the event of anyone being subjected to violence at work will be the key to resolving possible confrontations, and will be a deciding factor in the management of such situations.

STUDY QUESTION

14. What strategies are available to avoid the risk of violence at work?

(Suggested Answer is at the end.)

Substance Abuse at Work

IN THIS SECTION...

- Substance misuse can have various consequences on the health and safety of workers.
- Drugs and alcohol must be controlled by the employer through awareness of warning signs, consulting employees, considering safety-critical work, developing policy and screening, and support for affected workers.

Risks to Health and Safety from Substance Abuse at Work

DEFINITIONS

ALCOHOL

An addictive, narcotic drug that significantly impairs the senses and reaction times, even at low doses.

DRUG

A very broad term applied to a range of substances.

Here, the concern is prescription drugs (strong painkillers) and controlled drugs (illegal drugs such as cocaine). Some prescription drugs and most controlled drugs are addictive.

The effects of alcohol, drugs and solvents on the human body can lead to a number of related problems, namely:

- Alteration of personality.
- Reduced reactions.
- Lack of awareness.
- A change in attitude to danger.
- Hallucinations and blurred vision.
- Belligerence or extreme friendliness.
- Reduced efficiency, absenteeism.
- Dishonesty and theft.
- Poor time-keeping and misconduct.

The health and safety of drinkers and drug-takers themselves and others could be compromised by these possible effects, e.g. increased risk of accidents when working with machinery, hand tools or driving (a criminal offence).

Managing Substance Abuse at Work

The HSE suggests the following approach for managing substance abuse at work.

What the Issues Are and What to Look Out for

Misuse is not the same thing as dependence. Drug and alcohol misuse is the use of illegal drugs and misuse of alcohol, medicines and substances such as solvents.

Consider these warning signs, which could indicate drug or alcohol misuse:

- unexplained or frequent absences;
- a change in behaviour;

Drug and alcohol testing programmes may be put in place

13.6 Substance Abuse at Work

- unexplained dips in productivity;
- more accidents or near misses;
- performance or conduct issues.

Consult Employees

You must consult employees or their representatives on health and safety matters. Consultation involves you not only giving information to employees but also listening to them and taking account of what they say.

Consider Safety-Critical Work

Think about the kind of work you do and any safety-critical elements where drug or alcohol misuse could have a serious outcome, for example:

- using machinery;
- using electrical equipment or ladders;
- driving or operating heavy lifting equipment.

Where employees in safety-critical jobs seek help for alcohol or drug misuse, it may be necessary to transfer them to other work, at least temporarily.

Develop a Policy

All organisations can benefit from an agreed policy on drug/alcohol misuse. You could include a drug and alcohol policy as part of your overall health and safety policy. If an employee tells you they have a drug or alcohol problem, an effective policy should aim to help and support them rather than lead to dismissal.

Any drug and alcohol testing policy must be justified and clearly explained to workers. There are legal and ethical issues associated with testing regimes that must be carefully considered.

Screening

Some employers have adopted screening as part of their drug and alcohol policy. If you want to do the same, think carefully about what you want screening to achieve and what you will do with the information it gives you.

Supporting Employees with a Substance Abuse Problem

Support can be offered by:

- Training and awareness.
- Briefing managers and supervisors so they are clear about:
    - How to recognise the signs of drug or alcohol misuse.
    - The organisation's rules on drug and alcohol misuse.
    - What to do if they suspect an employee is misusing drugs or alcohol.
    - What to do when an employee tells them about a drug or alcohol problem.
- Offering support for employees:
    - Employees who misuse drugs or alcohol may ask for help at work if they are sure their misuse will be dealt with discreetly and confidentially.
    - Someone who is misusing drugs or alcohol has the same rights to confidentiality and support as they would if they had any other medical or psychological condition.
- Encouraging them to get help from their GP or a specialist drug or alcohol agency.
- Considering allowing someone time off to get expert help.
- Health advice and information.

Substance Abuse at Work — 13.6

Make sure there is information at work about where they can go for advice and help if they're concerned about drug or alcohol misuse.

STUDY QUESTION

15. What symptoms might an employer notice in an employee who is misusing drugs or alcohol?

(Suggested Answer is at the end.)

Summary

This element has dealt with some of the health hazards and controls relevant to physical and psychological ill health.

In particular, this element has:

- Outlined the health effects of exposure to noise, including the physical and psychological effects.
- Defined commonly used terms in noise measurement.
- Outlined the need to assess exposure to noise and the control limits.
- Outlined steps to control noise exposure by isolation, absorption, insulation, damping and silencing; and the importance of health surveillance.
- Outlined the effects of exposure to excessive vibration and the exposure standards which exist.
- Described the available vibration control measures, with reference to selection of equipment, maintenance and limiting exposure (including PPE).
- Outlined the importance of, and procedure for, health surveillance in regards to vibration.
- Described types of ionising and non-ionising radiation, including radon and artificial optical radiation, and their health effects.
- Outlined the methods and control of exposure for ionising and non-ionising radiation, including general radiation protection strategies and the role of health surveillance and monitoring.
- Outlined the workplace causes and effects of mental ill-health at work and preventive measures.
- Outlined the hazards and appropriate control measures for violence at work.
- Explained the hazards and appropriate control measures for substance misuse at work.

Exam Skills

Question

Scenario

You are currently reviewing the toolbox talks for powered hand-held equipment on site and you notice there is nothing covering the effects or precautions on Hand-Arm Vibration Syndrome (HAVS) in the toolbox talk library. You decide this needs to be rectified as a matter of urgency.

Task: Work Equipment

Back in the office you decide to prepare a briefing document to be used as a toolbox talk on the subject of HAVS. In the briefing document you need to create three clear headings:

(a) Ill-health effects associated with HAVS. **(2 marks)**

(b) What factors need to be considered when assessing the risks from HAVS. **(4 marks)**

(c) What precautions could be taken in order to help reduce the risk to the workers. **(4 marks)**

(Total: 10 marks)

Approaching the Question

Now think about the steps you would take to answer this question:

Step 1 The first step is to **read the scenario carefully**. Note the question is focussing on a particular hazard, the effects of it and what can be done to mitigate the risk.

You decide to create a briefing document for a toolbox talk so you will need to structure your approach using the three headings "Ill-health effects", "Factors to be considered in a HAVS assessment" and "Control measures to reduce the risk".

Step 2 Now look at the **task** - prepare some notes under the three headings in step 1.

Step 3 Next, consider the **marks** available. In this task, there are 2 marks available for the first part and 4 marks for the second and third part of the question. Tasks that are multi-part are often easier to answer because there are additional signposts in the question to keep you on track. In this task, you have to create a briefing document that is easy to understand, giving examples for each part can aid understanding. You will need to provide around 10 or more different pieces of information including examples for this task. The headings will allow you to keep your response separate – this will also help the examiner when marking.

Step 4 **Read the scenario and task again** to make sure you understand the requirements and ensure you have a clear understanding of hand-arm vibration syndrome. (Re-read your study text if you need to.)

Step 5 The next stage is to **develop a plan** - there are various ways to do this. Creating a bullet point list could be one way.

Exam Skills

Suggested Answer Outline

Ill-health effects:

- Vibration white finger.
- Nerve damage.

Factors to be considered in a HAVS assessment:

- Equipment likely to cause vibration and places of use.
- The employees and the magnitude, type and duration of exposure.
- Manufacturers information and vibration data.
- Specific working conditions, e.g. low temperatures.

Control measures to reduce the risk:

- Choice of equipment.
- Replace the tool or equipment for one with less vibration.
- Support the equipment allowing the operator to reduce grip.
- Anti-vibration mounts to isolate the operator from the vibration source.
- Insulating gloves.

Now have a go at the question yourself.

Example of How the Question Could be Answered

(a) *The ill-health effects from HAVS are vibration white finger, where the blood supply to the fingers shuts down and the fingers turn white. This is made worse by cold or wet conditions. Another form of harm is nerve damage to the fingers where the nerves stop working properly, resulting in a loss of pressure, heat and pain sensitivity.*

(b) *In the workplace an inventory of equipment held should be consulted to identify equipment that creates high levels of vibration in use. This could be done by using vibration instrumentation to establish vibration values. Identifying personnel who use equipment by establishing the magnitude of the vibration and the duration of exposure. Manufacturers by law have to indicate the vibration values on the equipment they produce. This however may increase with age of the equipment and how well maintained it is. Also look at the conditions the equipment is being used in. Low temperatures as well as wet conditions can increase the risk of personnel developing HAVS.*

(c) *In order to reduce the risk from HAVS, several control measures can be taken such as mechanising the activity – using a concrete breaker mounted on an excavator arm rather than hand operated. Change the equipment for one with less vibration generation characteristics. Support the equipment on tensioners or balancers allowing the operator to reduce grip and force. The fitting of anti-vibration mounts to isolate the operator from the vibration source. Finally, if the equipment is being used in a wet or cold environment the wearing of insulating gloves to keep the hands warm and dry will also reduce the effects of HAVS.*

Reasons for Poor Marks Achieved by Exam Candidates

- Not following a structured approach for the briefing document; failing to provide information on the three subject areas.
- Not expanding the answer beyond a few words as opposed to giving a sentence of explanation.

Health and Safety Management for Construction (UK)

Unit CN1 Final Reminders

Now that you have worked your way through the course material, this section contains some reminders to help you prepare for your NEBOSH open-book exam. It summarises the advice on how to approach your revision and the exam itself and has some hints and tips.

Unit CN1 Final Reminders

Preparing for the Exam

Open-book exams require advance planning. As you work through your studies, it's important to familiarise yourself with the study materials so that, at the time of the exam, you can find what you need quickly. RRC's course materials cover all the syllabus topics but to ensure that you do well in the exam, we recommend doing some additional reading. The study text provides some useful links to external sources - have a look at the '**More...**' boxes within the materials, these contain useful links to relevant topics.

> **MORE...**
>
> Further information can be accessed here:
>
> www.hse.gov.uk/pubns/books/l153.htm

'More...' boxes provide relevant links to further reading material (taken from RRC's CN1 Study Text)

At the time of the exam, you should not be reading information from your course materials for the first time or even re-reading the study text, you will simply run out of time. Being familiar with the materials will give you more time to concentrate on the scenario and less on frantic searching!

Don't forget that the normal requirements of an invigilated exam don't apply, so you can highlight and annotate your materials to help you locate topics easily and use your notes on the day. The Open University webpage has some great tips on highlighting and annotating materials for revision purposes, and can be accessed at: https://help.open.ac.uk/highlighting-and-annotating.

Keep your notes organised in advance; by doing so, you will be able to easily identify the relevant parts to compose your answers. This will ensure you optimise your time during the exam.

Revising for the Exam

One of the most common misconceptions about open-book exams is that there's no need to revise for them. In fact, you should study for them just as you would for any other exam! You won't be asked to recall information in the same way as for a closed-book exam but you still need the knowledge in order to apply it effectively and you need to be able to demonstrate that you have met the learning outcomes. Remember, the exam presents you with a problem in the form of a scenario to which you will need to give a solution, so you will need to use your knowledge and apply it to solve the problem.

Revision Tips

There is some useful information in your CN1 Study Text on how to tackle revision so here is a reminder of the main points, together with some additional advice.

Using the RRC Course Material

Read through all of the topics multiple times. This might be done by skimming over all of the content of CN1 to get a feel for structure and topics, followed by a more thorough read-through that jumps over the most complex topic areas, then a detailed read where you attempt to crack the complex topics.

Remember that understanding the information, and being able to remember and recall it, are two different skills. As you read the course material, you should **understand** it. In the exam, you have to be able to **remember**, **recall** and **apply** it. To do this successfully, most people have to go back over the material repeatedly.

Check your basic knowledge of the content of each element by reading the element Summary. The Summary should help you recall the ideas contained in the text. If it does not, then you may need to re-visit the appropriate sections of the element.

Using the Syllabus Guide

Download a copy of the NEBOSH Guide to the course, which contains the syllabus, from the NEBOSH website.

Map your level of knowledge and recall against Elements 1-13 in the syllabus guide. Look at the content listed for each element in the guide. Ask yourself the following question:

'If there is a question in the exam about that topic, could I answer it?'

You can even score your current level of knowledge for each topic in CN1 of the syllabus guide and then use your scores as an indication of your personal strengths and weaknesses. For example, if you scored yourself 5 out of 5 for a topic in Element 1, then obviously you don't have much work to do on that subject as you approach the exam. But if you scored yourself 2 out of 5 for a topic in Element 3 then you have identified an area of weakness. Having identified your strengths and weaknesses in this way, you can use this information to decide on the topic areas that you need to concentrate on as you revise for the exam.

You can also annotate or highlight sections of the text that you think are important.

Another way of using the syllabus guide is as an active revision aid:

- Pick a topic at random from any of the CN1 elements.
- Write down as many facts and ideas that you can recall that are relevant to that particular topic. Go back to your course material and see what you missed, and fill in the missing areas.

Setting Up for the Exam Day

To increase your chances of success and improve your confidence, we strongly advise that you complete a mock exam and get feedback from your tutor to prepare you for the real exam.

It is also recommended that you study in the same room and environment where you will carry out the exam, to ensure you are comfortable and set for the exam.

There are some things you can do to ensure you have the best possible set-up for the day:

- Make sure you can sit comfortably so that you are not distracted by uncomfortable posture.
- Ensure good lighting and a comfortable temperature.
- Know where your study materials are so that you spend less time looking for them at the time of the exam. Have your study materials within easy reach.
- If you live with anyone, make sure they are aware of when you are taking an exam to avoid unnecessary interruptions and distractions. Placing a friendly sign on your door may be a useful reminder for them!
- Switch off your phone, television and any other devices that may distract you.
- Have water and snacks handy.
- Ensure you can keep your computer or other device charged up.
- If you can't take the exam at home, book a quiet room with good lighting, charging point and Internet connection.

Unit CN1 Final Reminders

What to Do on the Day

On the day of the exam, you will be able to access the exam from 11.00 am UK time by logging in to the NEBOSH platform and downloading the file. You will have 48 hours to do the exam, starting from when the exam paper becomes available. This does not mean it should take you 48 hours to do the exam, nor does it mean that you have to be working for all that time; the 48-hour window is designed to allow time for you to read and analyse the exam questions, access your course materials, plan your answers, complete and submit the assessment, as well as take necessary breaks and fulfil your other everyday commitments. The paper should take around 8 hours to complete so make sure you are aware of the time.

So, how do you best utilise this time?

This is when your planning, studying and hard work will pay off. You will have your materials ready so you will be set up for a strong start.

You are not expected to write more than the current 4,500 words in total. You are allowed a 10% margin - you will not gain marks for going beyond this, so your answers should be relevant, cear and focused. RRC would strongly advise that you do not write more than 4,900 words in case examiners choose not to read beyond 4,950 words and you therefore miss out on marks.

Use your time wisely: work at your own pace but don't leave everything until the last minute. Review your materials, draft up your answers and allow time to make amendments. Take time to read the exam questions carefully. Refer to your prepared materials and notes with the following in mind:

- Your work should be your own, in other words do not copy content without referencing the source or this counts as plagiarism (more on plagiarism later).
- Do not communicate with anyone about the assessment.
- Do not ask or allow anyone to proof-read or help you with your work.

Another common mistake when doing an open-book exam is to refer to as many materials as possible; don't fall into this trap, you must be selective! Use only the materials that you need - again, this is why preparing them in advance is so important!

Don't become over-reliant on materials either, you must apply your own knowledge and argument. You want the materials to support your answer, not take over!

For details on how to download and submit your exam, please read NEBOSH's Technical Learner Guide on the NEBOSH website.

Approaching the Exam Questions

The open-book exam will test you on your ability to "demonstrate analytical, evaluation and creative skills as well as critical thinking" and how you apply your learning to your answers. In other words, you will need to show what you can do with your knowledge to solve the problems presented to you – and this may take practice.

The following example scenario and suggested answer illustrate how to approach the question, so make sure you read this section carefully.

Unit CN1 Final Reminders

Example Scenario

You are a newly appointed health and safety officer for a construction company. The company is currently engaged in groundwork operations in preparation for a multi-storey business complex. There are three tracked excavators operating on the site with 20 operatives carrying out various activities. One of the employees, an operative, has reported a minor incident. A reversing excavator has struck another vehicle in the staff car park on-site causing very minor damage.

The operative has reported the incident to the construction company site manager. The manager's minimal investigation found that the operator of the excavator was to blame, because insufficient attention was paid when the excavator was reversing.

The excavator operator was advised to be more careful in the future or disciplinary action would be taken.

At the next health and safety committee for the construction site you are discussing incident statistics and incident investigations that have taken place. Some of the committee members (worker representatives mostly) discredit the investigation into the reversing vehicle incident.

Task

How would you convince the other committee members to reopen the investigation?   **(10 marks)**

How to Answer the Question

Familiarise yourself with the study text materials/notes/flashcards/mind maps and your notes on incident investigation. Your notes will guide you to other information on incident investigation such as HSG245 *Investigating Accidents and Incidents*. This may be a PDF document on your desktop, or a downloaded version with highlighted sections from where you have looked at this document before. Try to visualise the scenario. You can use your web browser to research any terms that you are unfamiliar with.

Read the information provided in the scenario carefully. Consider that the information provided is all relevant and there are key indicators given to direct your answer. A key piece of information is that the investigation was "minimal". Your research will have indicated that a minimal investigation should be used for unlikely or rare occurrences with the potential for only minor injury. However, the worst possible event involving a reversing excavator, would be a fatality. You are advised that worker representatives "discredit the investigation" and from a minimal investigation it's likely that the manager only looked for immediate causes. This is consistent with the worker being blamed and no other causes being followed up. You are asked how you would convince the "other" committee members to reopen the investigation. The "other" members are likely to be management or employer representatives because the worker representatives want the investigation reopened. From your studies you will know that persuasive justification for managing safety can be covered under moral, legal, and financial reasons.

Unit CN1 Final Reminders

You could now consider an 'answer plan'.

- Incident investigation:
    - Purpose of investigation.
    - Immediate causes.
    - Underlying causes.
    - Root causes.
- Examples of immediate/underlying and root causes.
- How to take the investigation further - '5 whys'.
- Persuasive reasons to reopen the investigation:
    - Moral.
    - Legal.
    - Financial.
- Examples of moral, legal and financial reasons.

Suggested Answer

Remember to relate this to the scenario!

The committee members may have little or no experience on investigating accidents. I would advise them that the immediate cause of an incident would be the unsafe acts or unsafe conditions that led to the reversing vehicle striking the other vehicle. A medium or high-level investigation should establish why this happened.

The committee could be given examples of unsafe acts and conditions that a medium or high-level investigation may have revealed, e.g. damage to the excavators mirrors, faulty reversing camera or lack of a reversing assistant. Unsafe conditions may include poor levels of lighting or reversing too quickly.

A '5 whys' analysis could be used to further establish that the underlying causes themselves were caused by management system failings such as inadequate training of the worker. These are known as 'root causes' and without establishing the root causes, the incident may be repeated with another excavator and another plant operator - but this time it may be pedestrians are involved and the construction company would be investigating a fatality. This is morally and legally unacceptable.

I would advise that root causes may be organisational failures, job-related matters or personal factors. We may even find that we have not complied with legislation due to poor maintenance of the vehicle or lack of management systems of work. I would also advise committee members that monitoring safe systems of work was a legal duty that required a more complete incident investigation to take place.

Without a clear identification of the causes of the incident corrective action to prevent a recurrence would not be taken. I could go further and explain that the purpose of investigation is not to find someone to blame but to prevent a recurrence. This cannot be achieved without the full commitment of everyone in the construction company.

This answer has 295 words which is within 10% of the target set of our assumed maximum of 300 words for a 10-mark question - so we can keep within the required word count. As you complete your answer, refer to the documents and materials you have assembled to remind yourself of incident investigation, and be careful not to copy anything from the materials you have used without referencing it. Keep an eye on the word count to make sure you don't go over the allocated amount of words.

Example of an Insufficient Answer

Investigating accidents and incidents explains why you need to carry out investigations and takes you through 4 steps of the process:

- *Gathering information.*
- *Analysing the information.*
- *Identifying risk control measures.*
- *The action plan and its implementation.*

Gathering information is about the where and who of the event. Photographs and interviews are ways to gather information. Analysis of the information is what happened and why it happened. Human error should be considered at this stage. Then you need to identify control measures to prevent a recurrence. When you have identified control measures, create an action plan using the SMART planning technique.

Immediate cause: the most obvious reason why an adverse event happens, e.g. the guard is missing, the employee slips, etc. There may be several immediate causes identified in any one adverse event. Immediate causes include:

- *Inadequate safety devices.*
- *Poor housekeeping.*
- *Operator error.*
- *Wearing unsuitable footwear.*

The root causes are also known as management, planning or organisational failings.

There are three types of safety cultures: blame, no blame, and a just culture. It's clear this organisation has a blame culture and attempting to apportion blame is counterproductive, people become defensive and unco-operative. Only after a full investigation, not a minimal investigation, should individuals be blamed.

This answer has not used the scenario supplied and is far too general in its approach to attract many marks. It also lacks substance and detail. The explanation given would not encourage the reopening of the investigations because it has no moral or legal persuasion to convince the safety committee. Bullet-pointed lists do not provide sufficient evidence of your knowledge. The examiner will only have the words you have used to allocate marks against. What does 'poor housekeeping' mean related to the scenario? The examiner cannot guess what you mean and award marks against what it's thought you mean. You must be clear in your explanation, so your knowledge is demonstrated. This answer also uses acronyms, e.g. 'SMART'. Using an acronym seldom demonstrates knowledge, it is a far better response to write the words out fully before using the acronym. It would also be clear that significant sections of the answer are plagiarised from documents produced by the HSE. If the examiner investigated some of the phrases used, and found they reproduce someone else's work, further assessment would take place and a malpractice investigation would be conducted.

Note: The same scenario would be used to ask questions related to safety culture, reporting requirements, active/reactive monitoring, etc.

Risk Assessment Question

As should be clear by now, there is only one assessment in this course: an open-book exam featuring a detailed scenario, and questions that relate back to it. The assessment criteria for Elements 4 - 13 states "Produce a risk assessment that considers a wide range of construction hazards". The way this is handled in the open-book exam is by requiring you to conduct a risk assessment using the information provided in the scenario.

The CN1 open book exam has a risk assessment question that is worth 50 marks. This is a considerable number of marks, so it is essential that you do well in this question in order to successfully pass the exam. The word count does not apply to the Risk Assessment.

In your exam you will see that NEBOSH have identified a particular area of concern related to the scenario. This may be an excavation, a crane operation, working at height or any one of the hazards covered by Elements 4 through 13. Read the instructions carefully - NEBOSH may well focus the risk assessment on a particular aspect of a task given in the scenario. Instructions may tell you to NOT include a particular hazard related to the task. If this is the case, mentioning things you are instructed to ignore will not gain any marks. You will be presented with an answer template in the form of a table in your OBE. The table will have 3 columns labeled as 'a', 'b', and 'c'.

Column 'a' will have a heading of specific hazards. The scenario will have mentioned some specific hazards. In the sample paper a specific hazard related to the excavation is 'flooding'. Your OBE will say 'specific hazards **(10)**'. The number in brackets tells you 10 marks are available if you mention 10 specific hazards related to the task being assessed.

Column 'b' will have a heading of existing control measures. The scenario will have mentioned ALL these existing control measures. In the sample paper, an existing control measure is that the excavation is inspected every day before works starts. Using this information in column 'b' would gain one mark. Your OBE will say 'Existing control measures **(10)**'. The number in brackets tells you 10 marks are available if you mention 10 existing control measures related to the task being assessed.

Column 'c' will have a heading of 'additional control measures'. These additional measures will NOT be referenced in the scenario. You must use your knowledge (and any reference information you read) on Elements 1 through 13 to identify suitable controls that could be used to reduce the risk in the task being evaluated. Your OBE will say 'Additional control measures **(20)**'. The number in brackets tells you 20 marks are available if you mention additional control measures related to the task being assessed. If NEBOSH advise that 20 marks are available for additional control measures – it is safe to assume there are at least 20 and possibly 30 or more additional control measures that would attract a mark.

A useful tip for this section is to consider the elements of a 'safe system of work' – people, equipment, materials, environment.

As an example, for 'people' we may say they need to be trained – if the scenario tells you they are trained then this becomes an existing control measure. An additional control measure could be 'refresher training'.

For 'equipment' we may say it needs to be inspected before use – if the scenario tells you the equipment is being inspected before use this becomes an existing control measure. An additional control measure could be 'formal inspection by a competent person following a repair'.

Remember there will be at least 20 possible additional control measures so at least 20 additional control measures are needed to attract these marks!

The risk assessment section also has 10 marks available for prioritising **FIVE** of your additional control measures from question 10 (c). For **EACH** prioritised control, briefly justify your prioritisation in terms of risk reduction. When selecting priority risk control measures, consider the ones that are LIKELY to save life, or the controls that significantly reduce risk – or controls that are simple and cost effective to introduce. If you have identified 20 additional controls – what you now are being asked to do is to pick the five you think are most important to adopt.

NEBOSH Sample Paper

Now have a go yourself, you can access the CN1 sample paper at:
https://www.nebosh.org.uk/qualifications/health-and-safety-management-for-construction-uk/#resources

Plagiarism and Malpractice

You should follow the instructions and adhere to the guidance on the open-book exam. The answers that you submit must be your own. Any cases of suspected plagiarism will be investigated and any breaches will be dealt with in line with NEBOSH's malpractice policy which you can find on the NEBOSH website.

You must ensure that what you submit is your own work and if you quote or paraphrase anyone else's work, this must be referenced or it would constitute plagiarism.

The following counts as plagiarism:

- Inserting another author's sentences, paragraphs and ideas without referencing them, whether these are published or unpublished.
- Paraphrasing another author's work without referencing them.
- Collaborating with someone else (e.g. another learner) and submitting work that is either identical or very similar to theirs while claiming it was your own work.
- Paying someone to complete the work for you and submitting it as your own.
- Impersonation - when you ask someone else to complete the work for you and you pass it off as your own.

Your open-book exam will be marked by a NEBOSH examiner and will be scrutinised for plagiarism.

When taking a non-invigilated open-book exam you will need to declare that your submission is your own work and that you have not received help from anyone else. You will need to confirm you have read, understood and abided by NEBOSH's rules, by signing a Declaration.

Please note that NEBOSH reserves the right to submit your assessment to a plagiarism detection software package.

References

NEBOSH Open Book Examinations: Learner Guide - Guidance for preparing for an open book examination, NEBOSH, 2021

https://www.nebosh.org.uk/documents/open-book-examination-learner-guide

NEBOSH Health and Safety Management for Construction (UK) - Unit CN1: Managing Construction Safely, RRC Study Text, 2022

Note-taking techniques, The Open University, 2020

https://help.open.ac.uk/highlighting-and-annotating

Health and Safety Management for Construction (UK) - Qualification guide for learners, NEBOSH, 2022

https://www.nebosh.org.uk/qualifications/health-and-safety-management-for-construction-uk/#resources

NEBOSH Open Book Examinations: Technical Learner Guide, NEBOSH, 2022

https://www.nebosh.org.uk/documents/open-book-examination-technical-learner-guide

Unit CN1 Final Reminders

HSG245 Investigating accidents and incidents, HSE, 2004

https://www.hse.gov.uk/pubns/hsg245.pdf

OBE Sample Paper CN1, NEBOSH, 2022

https://www.nebosh.org.uk/qualifications/health-and-safety-management-for-construction-uk/#resources

Policy and Procedures for Suspected Malpractice in Examinations and Assessments, NEBOSH, 2022

https://www.nebosh.org.uk/documents/malpractice-policy-v16/

Good luck!

Unit CN1

Suggested Answers: Part 2

No Peeking!

Once you have worked your way through the study questions in this book, use the suggested answers on the following pages to find out where you went wrong (and what you got right), and as a resource to improve your knowledge and question-answering technique.

Suggested Answers to Study Questions

Element 8: Musculoskeletal Health and Load Handling

Question 1

WRULD stands for 'work-related upper limb disorder' and refers to ill-health conditions affecting the upper limbs, particularly the soft connecting tissues, muscles and nerves of the hand, wrist, arm and shoulder.

WRULDs arise from the repetition of ordinary movements (e.g. gripping, twisting, reaching or moving), often in a forceful and awkward manner, without sufficient rest or recovery time.

Question 2

Activities that can cause MSDs or WRULDs include::

- **Digging** - using a shovel by hand. This requires bending, twisting and lifting a load often while stooping.
- **Kerb laying** - lifting heavy kerbstones usually in very poor, stooped or even kneeling positions, and laying them accurately in-line and at the same level.
- **Movement and fixing of plasterboard** - requires the lifting (usually from a flat position to upright) and carrying of a very wide and tall plasterboard panel, then locating and fixing it in place.
- **Placement and finishing of concrete slabs** - usually lifting from a flat position, carrying and laying again in a flat position, before butting correctly to the other laid slabs and levelling.
- **Bricklaying** - often involves a lot of repetition as the items are small, although not heavy. They require a lot of twisting and turning movement to pick up and then lay the brick, while laying the cement requires similar repetition and movement.
- **Erecting and dismantling scaffolds** - requires reaching to remove poles and components from a vehicle, turning and carrying to a location (often some distance), then handling, turning, twisting to locate and fix the pole.
- **Use of display screen equipment** - architects and designers, as well as office staff and site management, may require the use of computers, and may be in temporary accommodation on site.

(Note: only three were required.)

Question 3

The three main ill-health effects due to poor design of tasks and workstations include:

- Physical stress - resulting in injury or general fatigue (aches, pains, etc.), usually from poor posture and excessive demands on manual dexterity, but also in respect of exposure to excessive noise and vibration.
- Visual problems - often through excessive brightness or prolonged, concentrated work on small objects, either on a computer display screen, or in respect of components used in a work process, e.g. in a construction site drawing office.
- Mental stress - mainly through excessive demands of task performance, lack of control over working processes and poor organisational and physical environmental conditions.

Question 4

Glare occurs when one part of the visual field is much brighter than the average brightness to which the visual system is adapted. Direct interference with vision is known as disability glare. Where glare causes discomfort, annoyance, irritability or distraction, this is known as discomfort glare and is often related to symptoms of visual fatigue such as sore eyes and headaches.

Glare may occur:

- As a result of a light source being directly in the line of vision (e.g. the setting sun shining in through a vehicle windscreen, or a badly positioned floodlight on site).
- From the reflection of light off a polished surface, e.g. a mirror or computer's screen.

Suggested Answers to Study Questions

Question 5

The main causes of injury are:

- Failing to use a proper technique for lifting and/or moving the object(s) or load.
- Moving loads which are too heavy.
- Failing to grip the object(s) or load in a safe manner.
- Not wearing appropriate PPE.

Question 6

The characteristics of a load which constitute a hazard are its weight, size, shape, resistance to movement, rigidity or lack of it, position of its centre of gravity, presence or absence of handles, surface texture, stability of any contents and the contents themselves.

Question 7

The main hazards in the working environment are:

- Constraints on movement and posture.
- Conditions of floors and other surfaces.
- Variations in levels.
- Temperature and humidity.
- Strong air movements.
- Lighting conditions.

Question 8

Redesign of the task may include:

- Sequencing - adjusting the sequence of tasks in a process to minimise the number of operations involving lifting and carrying loads.
- Work routine - reducing repetitive operations to allow variation in movement and posture, by such means as introducing breaks, job rotation and providing ways in which workers can operate at their own pace, rather than keeping up with a machine or process.
- Using teams - sharing the load by using teams of workers to carry out the task.
- Mechanising or automating the task.

Suggested Answers to Study Questions

Question 9

(a) The most common hazard of forklift trucks is that, with their small wheels and particularly when their forks are raised while carrying a load, they may become unbalanced, resulting in them shedding their load or tipping over. Other hazards arise from the constant need to reverse the truck, obscured vision when the load is raised and using unsuitable trucks for the working environment.

(b) The main hazards associated with lifts and hoists are falls from a height (from a landing level, from the platform or with the platform) and being hit by materials falling from the platform. Other hazards include being struck by the platform or other moving parts, and being struck by external objects or structures while riding on the platform.

(c) The main hazards associated with cranes are the risk of them becoming unbalanced and toppling over, the arm of the crane swinging out of control or the load striking something while being moved horizontally or falling.

(d) The main hazards of sack trucks are overloading, instability of the load, tipping when moving over uneven ground or on slopes, and careless parking.

Question 10

(a) PPE requirements for pallet trucks include safety footwear, as well as gloves and aprons to protect while handling loads.

(b) PPE requirements for cranes include safety helmets, rigger gloves, safety footwear and Hi-Vis clothing.

Question 11

Under **LOLER**, thorough examination is required:

- Before it is put into service for the first time, including the first time at a new location.
- At least every 6 months.
- Following any incident or accident that might have stressed the equipment.
- Following any change in conditions of use which could affect the safety of the equipment.

Ropes and chains on equipment used to lift people are to be inspected every working day.

Element 9: Work Equipment

Question 1

Definitions are:

(a) Any machinery, appliance, apparatus, tool or installation for use at work, including assembly arranged and controlled to function as a whole; equipment provided by an employer; and tools brought in by employees.

(b) Any activity involving work equipment. Includes starting, stopping, programming, setting, transporting, repairing, modifying, maintaining, servicing and cleaning.

Question 2

The act of fixing the UKCA (UKNI) mark to a product (and signing a Declaration of Conformity) constitutes a declaration by the manufacturer that the product meets the requirements of all the standards which apply to it. This provides some assurance that it is safe when properly installed, maintained and used for its intended purpose.

Question 3

The location in which the work equipment is used must be assessed to take into account any risks from particular circumstances, e.g. electrically-powered equipment used in wet or flammable atmospheres.

Suitable and sufficient lighting, which takes account of the operations to be carried out, must be provided at any place where a person uses work equipment.

Additional lighting, over and above general lighting levels, may be needed:

- In respect of particular areas of the machinery (e.g. where access to dangerous parts is required).
- For particular operations, e.g. maintenance.
- Where dangerous operations may be carried out, e.g. excavation, demolition, or working in the dark.

Question 4

A programme of preventive maintenance is required in cases where the safety of a piece of work equipment depends upon the installation conditions, or where it is exposed to conditions liable to cause deterioration to a dangerous state. The programme will be based on regular inspection of the work equipment.

Question 5

Workers are required to receive training in the use of a particular piece of work equipment where it involves a specific risk to health and safety, in particular:

- Where they are required to use that equipment for any task.
- Where they are required to carry out any repairs, modifications, maintenance or servicing on that equipment.
- Where they are required to use that equipment in certain environments, e.g. confined spaces.

All persons who use work equipment should receive training on the equipment they are expected to use and be aware of the health and safety implications, potential risks and precautions.

- Young persons (those under the age of 18) should be given special consideration due to their inexperience and immaturity, and be closely supervised by a competent person.
- Managers, supervisors and maintenance staff must also receive adequate training in relation to health and safety, the use of the equipment and risks entailed in its use.

Suggested Answers to Study Questions

Question 6

The main characteristics of an interlocked guard are that:

- It must prevent movement of the dangerous parts when the hazard area is open.
- It must not allow access to the hazardous area until the potential hazard has been made safe.
- It must not allow the machinery to operate until the guarding system is fully operational.

Question 7

(a) The risks in the use of hand tools arise from operator error, misuse and improper maintenance.

(b) The additional risks of portable power tools arise from the presence of the power source (and especially the cables) and the speed and force of the tool itself.

Question 8

The defective or damaged equipment should be taken out of service and quarantined with a label indicating the item should not be used until repaired - operators should be instructed never to use damaged or defective equipment. Visual checks of equipment should be carried out before use and defective items should be withdrawn.

Inspection and any subsequent tests and repairs should be carried out by a competent person and a record of inspection should be made and kept for the life of the equipment.

Question 9

There are six general factors about machines and the way in which people come into contact with them which cause specific **mechanical hazards**:

- Shape of the machine, e.g. whether it has sharp edges, angular parts, etc., which may be a hazard even if not moving.
- Relative motion of machine parts in relation to a person.
- Mass and stability of the machine or parts of it, including the workpiece.
- Acceleration of moving parts of a machine (or the workpiece), either under normal conditions or if something breaks.
- Inadequate mechanical strength of a machine or part of it.
- Potential energy of elastic components which may be translated into movement.

Question 10

Drawing-in injuries occur where a part of the body is caught between two moving parts and drawn into the machine, e.g. at 'in-running nips' where a chain and drive sprocket meet.

Question 11

Non-mechanical machinery hazards include: electricity; noise; vibration; hazardous substances; ionising radiation; non-ionising radiation; extreme temperatures; ergonomics; slips, trips and falls; and fire and explosion.

Question 12

Additional precautions include: warning notices, buoyancy aids, safety boats, platforms and gangways, ladders, good housekeeping, illumination of the water surface, consulting the weather forecast, first-aid facilities, protective clothing and equipment, life jackets, lifebuoys, rescue lines, safe operating procedures, safety nets and emergency procedures.

Question 13

Types of protective clothing and equipment that can be worn are:

- Safety helmets - these must be worn at all times, as anyone struck on the head and then falling into water is at a particular risk of drowning.
- Footwear - this should have non-slip soles. Rubber and/or thigh boots should be avoided due to them filling with water, which could result in the wearer being dragged under water.
- Safety harnesses and belts - these are permitted under the **Work at Height Regulations 2005** where it is not possible to provide a standard working platform or safety net, provided that they are always worn and always secured to a safe anchorage. There are a number of types, e.g. chest harnesses, full-body harnesses, safety rescue harnesses, etc. which are required to be properly selected for a particular use and the operatives trained and instructed in their use.

Element 10: Electricity

Question 1
The main effects of electricity on the body are ventricular fibrillation, cardiac arrest, extreme muscle contractions, burns at contact points and deep tissue burns.

Question 2
Arcing is the electrical bridging, through air, of one conductor with a very high potential to another nearby, earthed conductor. If the arc is connected to a person, the victim may be subject to both a flame burn from the arc and electric shock from the current which passes through the body. There is also a danger of burns from ultraviolet radiation and radiated heat, even where the arc does not actually touch a person. Arcing can also provide a source of ignition for fire.

Question 3
Risks associated with electricity:

- Electric shock (electrocution).
- Electrical burns (direct and indirect).
- Ventricular Fibrillation (VF).
- Asphyxiation.
- Cardiac arrest.
- Secondary injuries resulting from falling from access platforms or ladders.

(Note: only three were required.)

Question 4
Protective factors used for electrical equipment include:

- Fuses - a weak link in the circuit.
- Circuit breaker - a mechanical switch which automatically opens when the circuit is overloaded.
- Earthing - a low resistance path to earth for fault current.
- Isolation - cutting the power.
- Reduced low voltage - so that less current flows during an electric shock accident.
- Residual Current Devices (RCDs) - sensitive and fast-acting trips.
- Double insulation - separating people from the conductors using two layers of insulation.

(Note: only six were required)

Question 5
Three conditions that need to be met to ensure a safe system of work are no live working and safe isolation, locating buried services and protecting against overhead cables.

Question 6
The first action in treating a victim for electric shock should be to break any continuing contact between the victim and the current.

Suggested Answers to Study Questions

Question 7

(a) The three types of work are:

- No scheduled work or passage of plant under the lines.
- Plant/equipment will pass under the lines.
- Work carried out beneath the lines.

(b) The precautions required are as follows.

No scheduled work or passage of plant under the lines:

- Ground-level barriers are used to prevent close approach (minimum distance 6 metres unless changed/altered by the owner). (It depends on the line voltage.)
- Barriers are conspicuously marked (red/white stripes, plastic flags or hazard bunting). They may be stout poles, fences, tension wire fences earthed at both ends, oil drums filled with rubble/concrete, an earth bank (less than 1 metre), solid timber baulks or concrete blocks.
- Lines of plastic flags/bunting (3 to 6 metres above ground level) may be used, care being taken during erecting to avoid contact with or approach near to the live conductors.
- Where mobile plant/equipment is used then the length of the overhanging part of the plant or jib needs to be taken into account.
- Barriers will be required on one or two sides, depending on access to the worksite.
- No storage is permitted between or under the overhead lines or barriers.

Plant/equipment will pass under the lines:

- A minimum number of defined, fenced, level and well maintained passageways of restricted width are used.
- Goal-posts made of timber or plastic pipe are erected at each end of the passageway (parallel to the power lines), which is suitably signed/marked, e.g. red/white stripes.
- Warning notices should be in place on the approaches to the crossings and regarding the height of the crossbar, instructing drivers to lower their jibs and keep below this height.
- Work after dark requires that any notices/crossbars should be adequately and suitably illuminated.
- Lighting should be at ground level directing the light upwards towards the conductors.

Work carried out beneath the lines:

- Additional precautions including erecting barriers, goal-posts and warning notices/signs may be required to prevent the upward movement of scaffold poles, crane jibs, excavators and buckets.
- If there is a risk to the health and safety of workers carrying scaffold poles, ladders or other conducting objects, then these and any mobile equipment should be excluded if possible, or shorter scaffold tubes, ladders or metal sheeting should be used.
- Mobile plant and equipment should be modified by physical restraints to limit their operations, e.g. mechanical stops, limit switches.
- A roof could be constructed over the work area to prevent contact with the live lines.
- The use of proximity warning devices and insulating guards without other safety precautions is regarded as unacceptable.
- Great care should be taken not to reduce any distances/height clearances during any type of construction work, e.g. dumping, tipping waste, landscaping, scaffolding, etc.
- For work where there are buried services, extra precautions will be required.

Suggested Answers to Study Questions

Question 8

Four cable locating devices are a hum detector, radio frequency detectors, transmitter-receiver detectors and metal detector.

Question 9

Precautions when cables are exposed are:

- With spans of more than 1m should be supported.
- Should not be used as handholds or footholds.
- Should not be moved unless absolutely necessary.
- Should be reinstated with advice from the owner.

Element 11: Fire

Question 1

The processes of heat transmission/fire spread are:

(a) Convection.

(b) Radiation.

(c) Conduction.

Question 2

Direct burning is not shown.

Question 3

(a) Friction is the process whereby heat is given off by two materials moving against one another. In the absence of a lubricant or cooling substance, it can result in the surfaces of the materials becoming hot or actually producing sparks, either of which may be sufficient to cause ignition. Friction can be caused by impact (one material striking another), rubbing (when moving parts of a machine contact stationary surfaces) or smearing (e.g. when a steel surface coated with a softer light metal is subjected to a high specific bearing pressure with sliding or grazing).

(b) A space heater is designed to give off considerable heat and, close to the heater, temperatures may be very high. Fire may be started by combustible materials being placed too close to the source of the heat (through radiation) or by them actually touching the hot surfaces of the heater itself.

Question 4

The three ways of extinguishing a fire are starvation (removing the fuel), smothering (removing the oxygen) and cooling (removing the heat).

Question 5

The classifications of fire are:

(a) Class C - fires involving gases or liquefied gases.

(b) Class B - fires involving flammable liquids or liquefied solids.

(c) Class A - fires involving solid, mainly carbonaceous, materials (here, most likely paper and furniture, etc.).

Question 6

Fire risk can be minimised by ensuring that wood shavings and dust are cleared regularly and ignition sources such as sparks from electrical equipment do not come into contact with combustible materials.

Question 7

The volume of flammable liquids in use at any one time should be minimised (up to 250 litres is usual) and it should be held in appropriate (usually metal), correctly labelled containers with secure lids. The need to decant highly flammable liquids from one container to another should be minimised, thereby reducing the risk of spillages. Storage areas should be well ventilated and drip trays and proper handling aids should be provided. A method for dealing with spillages and the disposal of empty containers and contaminated waste is required.

Question 8

Flammable liquids are categorised as flammable, highly flammable and extremely flammable.

Suggested Answers to Study Questions

Question 9

Good housekeeping prevents fire in the following ways:

- Combustible and flammable materials are regularly removed from work areas.
- Items that can't be removed are covered with fire-retardant blankets.
- Waste bins are emptied regularly.
- Site areas are regularly cleaned and kept free of litter and rubbish.
- Safe disposal of all waste materials is arranged. Site fires are banned.
- Skips are placed at least three metres from buildings and other structures.
- Pedestrian routes are always kept clear.

Question 10

(a) The fire resistance of timber depends on the 'Four Ts' – the thickness or cross-sectional area of the piece, the tightness of any joints involved, the type of wood and any treatment received.

(b) The fire resistance of reinforced concrete depends on the type of aggregate used and the thickness of concrete over the reinforcing rods.

(c) The fire resistance of a brick wall depends on its thickness, the applied rendering or plastering, whether the wall is load-bearing or not, and the presence of perforations or cavities within the bricks.

Question 11

The beam will distort, possibly causing the collapse of any structure it is supporting. It will also conduct heat and increase the possibility of fire spread.

Question 12

Flame-retardant paint, when exposed to heat, bubbles rather than burns, thereby giving additional protection to the covered timber.

Question 13

- Zone 0 - explosive atmosphere present continuously or for long periods.
- Zone 1 - explosive atmosphere likely to occur in normal operation.
- Zone 2 - explosive atmosphere not likely to occur in normal operation; if it does, it is only for a short time.

Question 14

Manual systems alone can only raise an alarm over a limited area and for a limited time. There needs to be some means for the person raising the alarm to make it more widespread - by using a phone or public address system, or a manual/electric system.

Question 15

Three fire detection methods include:

- Detection of smoke or other fumes by ionisation or optical smoke detectors.
- Detection of flames by ultraviolet and infrared radiation detectors.
- Detection of heat by fusion or expansion heat detectors.

Suggested Answers to Study Questions

Question 16

Classes of fire and suitable extinguishers are:

(a) Water - class A.

(b) Carbon dioxide gas - classes B and electrical fires.

(c) Dry powder - classes A, B, C and electrical. (There are special dry powders for class D.)

(d) Foam - class A and B. (Only specialist foams are suitable for some electrical fires.)

(e) Fire blankets - classes B and F.

Question 17

Under BS EN 3-10:2009 all extinguishers are now red, with colour identification on each, as follows:

- Water - white lettering.
- Carbon dioxide - black.
- Foam - cream.
- Dry powder (ABC) - blue.
- Dry powder (D) - violet.

Question 18

Fire extinguisher training should cover:

- General understanding of how extinguishers operate.
- The importance of using the correct extinguisher for different classes of fire.
- Practice in the use of different extinguishers.
- When and when not to tackle a fire.
- When to leave a fire that has not been extinguished.

Element 12: Chemical and Biological Agents

Question 1

The physical forms of chemical agents which may exist in the workplace are dusts, fibres, fumes, gases, mists, vapours and liquids.

Question 2

The five main health hazard classifications of chemicals are toxic, harmful, corrosive, irritant and carcinogenic.

Question 3

Mists are usually small liquid droplets (aerosol) suspended in the air, while fumes are fine solid particles which are created by condensation from a vapour, given off in a cloud.

A potential source of mists is where paint is being sprayed; fumes are emitted from welding processes.

Question 4

Acute ill-health effects arise where the quantity of a toxic or harmful substance absorbed into the body produces harmful effects very quickly, i.e. within seconds, minutes or hours. Chronic ill-health effects arise where the harmful effects of a substance absorbed into the body take a very long time to appear - months or perhaps years.

Question 5

The main routes of entry are inhalation, ingestion, absorption and injection. A further route is aspiration.

Question 6

Inhalable substances are capable of entering the mouth, nose and upper reaches of the respiratory tract during breathing. Respirable substances are capable of deeper penetration to the lung itself. They are generally 7 microns or less. It is the size of the individual particle that determines whether a substance such as dust is inhalable or respirable.

Question 7

A product label must give the following information:

- Name, address and telephone number of the supplier.
- Nominal quantity of the substance/mixture (though this may be elsewhere on the package) - but only where made available to the general public.
- Product identifiers:
  - for substances: name and identification number (EC number, CAS number or inventory number);
  - for mixtures: trade name, and the identity of all the substances (maximum of 4) in the mixture which contribute to its classification.
- Hazard pictograms.
- Signal word (as applicable).
- Hazard statements (as applicable).
- Precautionary statements (as applicable).
- Supplementary information.

Suggested Answers to Study Questions

Question 8

Safety data sheets are intended to provide users with sufficient information about the hazards of a substance or preparation for them to take appropriate steps to ensure health and safety in the workplace in relation to all aspects of its use, including its handling, transport and disposal.

Question 9

In passive sampling devices, the air sample passes through/into the device by means of natural air currents and diffuses into a chamber containing an absorbent material which can be removed for later analysis. In active sampling devices, the air sample is forced through the instrument by means of a pump.

Question 10

The limitations of stain tube detectors are as follows:

- They provide a spot-sample for one moment in time rather than an average reading.
- They can have an accuracy of +/-25%, which is not particularly accurate.
- The correct number of strokes must be used; losing count and giving too few/too many will give inaccurate results.
- The volume of air sampled may not be accurate due to incorrect assembly interfering with the air flow (through leaks, etc.) or incorrect operation.
- There may be the possibility of cross-sensitivity of tube reagents to substances other than the one being analysed.
- There may be problems caused by variations in temperature and pressure.
- The indicating reagent in the tubes may deteriorate over time.
- There may be variations in the precise reagent make-up between tubes.

(Note: only three were required.)

Question 11

Smoke tubes are used to test the effectiveness of ventilation or air-conditioning systems and chimneys, to detect leaks in industrial equipment, to assess relative air pressures used in certain types of local ventilation system, and to provide general information about air movements in a work area.

Question 12

Guidance Note EH40 sets out the workplace exposure limits for substances hazardous to health.

Question 13

A Workplace Exposure Limit (WEL) is the maximum concentration of an airborne substance, averaged over a particular period, to which employees may be exposed by inhalation under any circumstances, as contained in the HSE Guidance Note EH40.

Question 14

WELs are expressed as time-weighted averages, meaning that measurements are taken over a particular time period (15 minutes for short-term limits or 8 hours for long-term limits) and then averaged out. The concept of time-weighted averages allows concentration levels to exceed the limit, provided that there are equivalent exposures below it to compensate.

Suggested Answers to Study Questions

Question 15

The limitations of WELs are as follows:

- They are designed only to control absorption into the body following inhalation.
- They take no account of human sensitivity or susceptibility (especially in relation to allergic response).
- They do not take account of the synergistic effects of mixtures of substances.
- They do not provide a clear distinction between 'safe' and 'dangerous' conditions.
- They cannot be applied directly to working periods which exceed 8 hours.
- They may be invalidated by changes in temperature, humidity or pressure.

(Note: only three were required.)

Question 16

The two reference periods are 15 minutes (for STELs) and 8 hours (for LTELs).

Question 17

The principles of control are:

(a) Substitution.

(b) Work process change.

(c) Reduced time exposure.

(d) Substitution.

Question 18

Local Exhaust Ventilation (LEV) is a control measure for dealing with contaminants generated from a point source. Dilution ventilation deals with contamination in the general atmosphere of a workplace area.

Question 19

Dead areas are areas in the workplace which, owing to the air-flow pattern produced by the positioning of extraction fans and the inlets for make-up air used in the ventilation system, remain dormant and so the air is not changed. This is a problem for dilution ventilation as the harmful contaminant remains in this area.

Question 20

The four main types of respirator are: filtering facepiece respirators; ori-nasal or half-mask respirators; full-face respirators; and powered visor respirators.

For breathing apparatus, the three main types are: fresh-air hoses, compressed airlines and self-contained systems.

Suggested Answers to Study Questions

Question 21

Key factors in the selection of the appropriate respirator:

- Contaminant concentration and its hazardous nature (e.g. harmful, toxic).
- Physical form of the substance (e.g. dust, gas, vapour).
- Level of protection offered by the RPE.
- Presence or absence of normal oxygen concentrations.
- Duration of time that it must be worn.
- Compatibility with other PPE that must be worn.
- Shape of the user's face and influences on fit.
- Facial hair might interfere with an effective seal.
- Physical requirements of the job, e.g. the need to move freely.
- Physical fitness of the wearer.

Question 22

The main purpose of routine health surveillance is to identify, at as early a stage as possible, any variations in the health of employees which may be related to working conditions.

Question 23

Six chemical agents from: petrochemicals; organic solvents; isocyanates; lead; silica; cement dust; asbestos; fibres; carbon dioxide; carbon monoxide; nitrogen.

Four biological agents from: blood-borne viruses; tetanus; leptospirosis (Weil's disease); Legionellosis (Legionnaires' Disease or Pontiac Fever); hepatitis.

Question 24

Gases described as 'asphyxiant' (e.g. carbon dioxide (CO_2) and carbon monoxide (CO)) do not cause direct injury to the respiratory tract when inhaled, but reduce the oxygen available to the body.

Question 25

Sources of organic solvents used in construction: paints, varnishes, adhesives, pesticides, paint removers and cleaning materials.

(Note: only three were required.)

Ill-health effects include irritation and inflammation of the skin, eyes and lungs, causing dermatitis, burns and breathing difficulties including occupational asthma and sensitisation. Vapours given off are usually flammable, and may be narcotic (e.g. toluene) progressively causing drowsiness, nausea and unconsciousness. Some organic solvents are carcinogenic.

Question 26

Controls used to avoid or reduce exposure to cement dust and wet cement:

- Eliminating or reducing exposure.
- Use of work clothing, and PPE such as gloves, dust masks and eye protection.
- Removal of contaminated clothing.
- Good hygiene and washing on skin contact.
- Health surveillance of skin condition to control chrome burns and dermatitis.

Suggested Answers to Study Questions

Question 27

The three main types of asbestos are:

- White (chrysotile).
- Blue (crocidolite).
- Brown (amosite).

Question 28

A procedure must be in place covering the actions to take on discovering asbestos in unknown locations:

- Stop work immediately.
- Prevent anyone entering the area.
- Arrangements should be made to contain the asbestos – seal the area.
- Put up warning signs – 'possible asbestos contamination'.
- Inform the site supervisor immediately.
- If contaminated, all clothing, equipment, etc. should be decontaminated and disposed of as hazardous waste.
- Undress, shower, wash hair; put on clean clothes.
- Contact a specialist surveyor or asbestos removal contractor.

Question 29

Sampling for asbestos in the air should be carried out by trained staff, in three situations:

- **Compliance sampling** - within control or action limits.
- **Background sampling** - before starting work (i.e. removal).
- **Clearance sampling** - after removal and cleaning the area.

Suggested Answers to Study Questions

Element 13: Physical and Psychological Health

Question 1

This refers to a daily personal exposure to noise ($L_{EP,d}$) at a level of 85 dB(A) over the course of a working day (8 hours), or an equivalent exposure over a shorter period. The significance of the level it relates to is that it is the upper exposure action level.

Question 2

Ear defenders can be uncomfortable when worn for long periods and users may be tempted to either remove or constantly adjust them. They may also be incompatible with other items worn such as spectacles or other items of PPE, and they must be routinely inspected, cleaned and maintained to ensure their effectiveness.

Ear plugs are difficult to see when properly fitted, leading to problems of supervision and enforcement, and workers not replacing them on a regular basis can lead to ear infections. In addition they are easy to lose or misplace.

There is a general limitation on the level of noise reduction that can be achieved, depending on the quality and type of ear protection. Taking off the protection reduces its effectiveness. In addition, the seal between the ear and the protective device may be less than perfect due to long hair, thick spectacle frames and jewellery, incorrect fitting of plugs or the wearing of helmets or face shields.

Question 3

Audiometry allows:

- Recognition of existing hearing loss (before starting employment).
- Further damage or hearing loss during employment to be identified.
- The removal or exclusion of workers from high noise areas (to protect from further loss).
- An evaluation of the effectiveness of noise controls.

(Note: only three were required.)

Question 4

Symptoms of hand-arm vibration syndrome include:

- Fingers turning white, before becoming red and painful when the blood supply returns (vibration white finger).
- Loss of pressure, heat and pain sensitivity (nerve damage).
- Reduction in grip strength and manual dexterity (muscle weakening).
- Abnormal bone growth at the finger joints (joint damage).

Suggested Answers to Study Questions

Question 5

There are a number of preventive and precautionary measures which can be taken in regards to tools and equipment:

- **Choice of equipment**:
  - Mechanise the activity - use a concrete breaker mounted on an excavator arm rather than hand-operated.
  - Change the tool or equipment for one with less vibration generation characteristics.
  - Use tools that create less vibration, e.g. a diamond-tipped masonry cutter instead of a tungsten hammer drill.
  - Support the tools (e.g. tensioners or balancers), allowing the operator to reduce grip and feed force.
  - Add anti-vibration mounts to isolate the operator from the vibration source.
- **Maintenance**:
  - Keep moving parts properly adjusted and lubricated.
  - Keep cutting tools sharp.
  - Replace vibration mounts before they wear too badly.
  - Ensure rotating parts are checked for balance.
  - Keep all equipment clean - especially look for corrosion.

Question 6

Regulation 7 of **CVAWR** requires that health surveillance should be conducted where appropriate, e.g. in cases where the risk assessment shows a risk of developing vibration-related conditions, or employee exposure reaching action levels. Records of this health surveillance should be kept.

Where an identifiable disease related to vibration exposure is discovered, monitoring is required to minimise the health effects and maintain adequate control.

Question 7

The non-ionising radiation types are:

(a) Radio frequency.

(b) Infrared radiation.

(c) Ultraviolet radiation.

(d) Visible radiation.

Question 8

Acute effects of exposure to high doses of ionising radiation include:

- Sickness and diarrhoea.
- Hair loss.
- Anaemia, due to red blood cell damage.
- Reduced immune system due to white blood cell damage.

All of the cells of the body are affected by the radiation, but some more than others. A large enough dose can kill in hours or days.

Suggested Answers to Study Questions

Question 9

Artificial optical radiation has effects such as:

- Burns or reddening of the skin.
- Burns or reddening of the surface of the eye (photokeratitis).
- Burns to the retina of the eye.
- Blue-light damage to the eye (photoretinitis).
- Damage to the lens of the eye that causes early cataracts.

Question 10

Radon is a naturally occurring radioactive gas originating from uranium, occurring naturally in rocks and soils - radon levels are much higher in certain parts of the UK. The highest levels are found in underground spaces such as basements, caves, mines, utility industry service ducts and in some areas in ground floor buildings, as they are usually at a higher pressure than the surrounding atmosphere. It usually gets into buildings through gaps and cracks in the floor.

All workplaces can be affected in radon-affected areas.

Question 11

Where controlled areas have been designated, the employer must appoint a qualified Radiation Protection Adviser (RPA). Usually from external organisations, RPAs must have particular experience of the type of work the employer undertakes and be able to provide advice and guidance on the following matters:

- Compliance with current legislation.
- Local rules and systems of work.
- Personnel monitoring, dosimetry and record-keeping.
- Room design, layout and shielding.
- Siting of equipment emitting ionising radiation.
- Siting and transport of radioactive materials.
- Leakage testing of sealed sources.
- Investigation of incidents, including spillages or losses.

(Note: only four were required.)

Suggested Answers to Study Questions

Question 12

The main causes of work-related stress include:

- **Demands**, in terms of workload, speed of work and deadlines - these should be reasonable and where possible, set in consultation with workers. Working hours and shift patterns should be carefully selected and flexible hours allowed where possible. Workers should be selected on their competence, skills and ability to cope with difficult or demanding work.
- **Control** - employees should be encouraged to have more say in how their work is carried out, e.g. in planning their work, making decisions about how it is completed and how problems will be tackled.
- **Support** - feedback to employees will improve performance and maintain motivation. All feedback should be positive, with the aim of bringing about improvement, even if this is challenging. Feedback should focus on behaviour, not on personality. Workers should have adequate training, information and instruction.
- **Relationships** - clear standards of conduct should be communicated to employees, with managers leading by example. The organisation should have policies in place to tackle misconduct, harassment and bullying.
- **Role** - an employee's role in the organisation should be defined by means of an up-to-date job description and clear work objectives and reporting responsibilities. If employees are uncertain about their job or the nature of the task to be undertaken, they should be encouraged to ask at an early stage.
- **Change** - if change has to take place, employees should be consulted about what the organisation wants to achieve and given the opportunity to comment, ask questions and get involved. They should be supported before, during and after the change.

(Note: only one example of a preventive measure was required for each.)

Question 13

Reducing occupational stress in construction work includes:

- Ensuring policies and procedures are in place to cover harassment, discrimination, violence and bullying.
- Having managers and supervisors trained to recognise the symptoms of stress.
- Reducing levels of noise on site.
- Ensuring adequate lighting in all work areas (especially at night).
- Ensuring adequate welfare and rest facilities are available on site.
- Ensuring high standards of housekeeping and maintenance of equipment and facilities.
- Discouraging long working hours.
- Introducing job rotation and increasing work variety as much as possible.
- Ensuring competence and ensuring people are properly matched to jobs.
- Maintaining high levels of communication and involving workers in making decisions.

Question 14

Strategies available to avoid the risk of violence at work include:

- Minimisation of cash handling, minimisation of customer/client frustration and refusing access to potentially violent customers and clients.
- Physical security measures such as secure doors, surveillance, improved lighting, etc.
- Reviewing systems of work to ensure greater security of staff, e.g. of lone workers.
- Training of employees who may find themselves in high-risk situations and the provision of information on the company policy and actions to take in the event of anyone being subjected to violence at work.

Question 15

The effect of alcohol and drugs on the human body can lead to a number of symptoms, namely alteration of personality, reduced reactions, lack of awareness, a change in attitude to danger, hallucinations, blurred vision, belligerence or extreme friendliness, reduced efficiency, absenteeism, dishonesty and theft, poor time-keeping and misconduct.